Andrew W. Hass is Reader in Religion at the University of Stirling and Executive General Editor of the journal *Literature and Theology*. He is the author of *Auden's O: The Loss of One's Sovereignty in the Making of Nothing* (2013) and *Poetics of Critique: The Interdisciplinarity of Textuality* (2003), and co-editor of *The Oxford Handbook of English Literature and Theology* (2007, with David Jasper and Elisabeth Jay).

'Andrew Hass' *Hegel and the Art of Negation* is a comprehensive and magisterial rethinking of Hegel, not only calling forth a truly contemporary Hegel, but a new Hegel: a Hegel at the very center of thinking as such, and of all thinking, or all truly critical and universal thinking.'

**Thomas J.J. Altizer, Professor Emeritus of Religious Studies,
The State University of New York, Stony Brook**

'This is an engaging and provocative exploration of the Hegelian art of negation. It offers us an attentive and insightful reading of the Hegel of negation, as well as of the negation of Hegel by many significant contemporary thinkers. It represents a fertile return to Hegel, a return of Hegel, and illuminates the central significance of the triad art/religion/philosophy. Its fresh consideration of the importance of this triad is especially worthy of commendation. Andrew Hass is well informed about Hegel himself as well as the latter's commentators and critics. Throughout one finds touches of rhetorical playfulness that serve to intrigue the reader. The book is important for trying to bring Hegel's view of art into proper communication with the full dimensions of Hegel's overall philosophical venture.'

**William Desmond, Professor of Philosophy, Katholieke Universteit Leuven
and David R. Cook Visiting Professor of Philosophy, Villanova University,
author of** *Hegel's God*

'Here is a truly artistic homage to the artistry at the heart of Hegel's thought: the sheer restless negativity that forever differentiates Hegel from 'Hegelianism', or perhaps Hegelianism from other '-isms'. In this enjoyable text Hegel returns – negating the negation he must, of course, also undergo – as freshly thought-provoking and challenging as ever.'

**Andrew Shanks, Canon Theologian, Manchester Cathedral,
author of** *Hegel's Political Theology* **and of** *Hegel and Religious Faith*

HEGEL AND THE ART OF NEGATION

negativity, creativity and contemporary thought

ANDREW W. HASS

Published in 2014 by I.B.Tauris & Co Ltd
6 Salem Road, London W2 4BU
175 Fifth Avenue, New York NY 10010
www.ibtauris.com

Copyright © 2014 Andrew W. Hass

Cover illustration: Portrait of Georg F. Hegel (1770–1831), German philosopher (photo by Time Life Pictures/Mansell/Getty Images)

The right of Andrew W. Hass to be identified as the author of this work has been asserted by him in accordance with the Copyright, Designs and Patents Act 1988.

All rights reserved. Except for brief quotations in a review, this book, or any part thereof, may not be reproduced, stored in or introduced into a retrieval system, or transmitted, in any form or by any means, electronic, mechanical, photocopying, recording or otherwise, without the prior written permission of the publisher.

Library of Modern Religion vol. 37

ISBN: 978 1 78076 557 0 (HB)
978 1 78076 558 7 (PB)

A full CIP record for this book is available from the British Library
A full CIP record is available from the Library of Congress

Library of Congress Catalog Card Number: available

Text design, typesetting and eBook by Tetragon, London
Set in 10.5/13.5 Sabon Monotype

For Tom Altizer, our Black Monk

Contents

Acknowledgements	ix
Introduction – Returning Anew	1
Part I: The Hegel of Negation	
Chapter 1	23
Negation's Art in *Phenomenology of Spirit*	
Chapter 2	39
Negation's Logic in *Science of Logic*	
Chapter 3	55
Art's Negation in *Aesthetics*	
Part II: The Negation of Hegel	
Chapter 4	77
The Returning of Hegel and Negation: Sartre and Hyppolite	
Chapter 5	91
The Tolling of Hegel and Negation: Derrida	
Chapter 6	113
The Living of Hegel and Negation: Kristeva, Nancy, Agamben, Žižek, Malabou	
Part III: Furthering Hegel	
Chapter 7	155
The Ought of Negation	
Conclusion: Art-Religion-Philosophy Re-formed	169
Notes	183
Bibliography of Works Cited	213
Index	223

Acknowledgements

This book is the culmination of a long process in rethinking Hegel that began with thinking upon, properly and perhaps for the first time, the idea of nothing. One soon realises that to make nothing or nothingness the object of thinking, one must first marshal into service a preposition – one thinks *around*, *about*, *upon* or *within* nothing – before one can *think nothing* directly, if at all. The preposition acts as a *pre-position* before the assumption of any direct position per se. I found that this pre-position became my previous book, *Auden's O: The Loss of One's Sovereignty in the Making of Nothing*, the writing of which, in the end, overlapped with this book, but which nevertheless still, in many ways, acts as its propaedeutic. Of course, this present text does not in itself achieve *thinking nothing* directly or immediately, a thinking which may ultimately be, as Parmenides taught us, impossible. At best it achieves the thinking *of* nothing, which remains prepositionally constructed, but now, as I will argue below, with the ambiguities of the genitive potently and productively in play. Hence the art *of* negation.

Even the thinking of nothing, as the art of negation, requires bold commitment, and I am extremely grateful to Alex Wright for his belief and immense patience in the process just described. I am also indebted to all my interlocutors around Hegel, living and passed on, many of whom feature in the argument to follow, and some of whom have helped situate previous thinking and iterations. Chapter 7, most notably, is a well reworked chapter first presented at Chicago's Divinity School in 2010 and published later in a volume on ethics, *Theological Reflection and the Pursuit of Ideals – Theology, Human Flourishing and Freedom*, edited by David Jasper and Dale Wright (Ashgate, 2013). In whatever context or discipline – and this book, like its predecessor, embraces interdisciplinarity unreservedly, out of a necessity inherent in its theme – we have all been learning and re-learning to *think Hegel* directly, which is necessarily a thinking *of* Hegel, and, by turn, his very negation.

INTRODUCTION

Returning Anew

Hegel is on the move again. Or, Hegel is again circling back on himself. Or, Hegel is making a beginning – again.

Each of these is a permutation of recurrence *within* Hegel's thought; but each is also a permutation of recurrence happening right now, *with* Hegel's thought. What accounts for these recurrences of Hegel? Why does Hegelian thought, which has already had such profound influence, keep resurging, and why a 'renaissance' now,[1] in this early millennium wracked by historical circumstances that seem so counter to the Hegelian understanding of dialectical historical progress?

Is it because the *within* and the *with* converge: the very content of Hegel's thought enacts itself in the form in which it is received and appropriated by subsequent generations? Or is it because this very convergence, as synthesis of form and content, is itself played out both within Hegel's thought and with Hegel's thought, in a multivalency of syntheses? Or is it that such multivalency is self-confirming, in the manner that Hegel's thought had determined: what happens *within* our individual thinking is precisely what develops or is made manifest *with* our collective thinking in the world?

Or is it because, to use Hegel's own terminology, the in-itself of *Hegel's* thought becomes the for-itself of *Hegelian* thought, which in turn becomes the in-and-for-itself of a *Hegelianism*, remade or refuted in contemporary terms?

Does such a cultural return to Hegel mean, in sum, a confirmation of Hegel's own thought? And does it mean this only by way of its dis-confirmation – moving on from, circling back on, beginning anew any kind of Hegelianism, once more?

All these questions, dense as they first appear, form the thematics of this book and its central question: what accounts for Hegel's continual recurrence, both now and always? And the permutations of that recurrence – moving, circling, making, beginning – form the essence of its argument: *Hegel keeps permuting (himself)*.

HEGEL VS. HEGELIANISM(S)

Let us make something clear from the outset: what is not making a return is Hegelianism in any traditional sense. For that Hegelianism does not and cannot by rights *return*. The tradition, even in all its variegation, has understood Hegelianism in terms of a proper *ism*, something systematic and doctrinal in character. Here, its character tends to unfold along general categories (despite the inherent risks of such generalisations, and for Hegel especially): the system of dialectics, and the doctrine of the Absolute.

In the system, a dialectical dynamic, one too often reduced to that triangulation of movement between thesis, antithesis and synthesis, enacts a forward-moving progression, a teleology, towards a world-historical culmination, in which the spirit of the age, Hegel's all-encompassing *Geist*, becomes fully self-conscious of itself as the spirit *of an age*, this great modern age, as it is materially and conceptually manifested in its greatest of cultural products: art, religion and philosophy. In this system, contrary and conflictual forces serve only to promote a higher reality, the third term of an ever-advancing treble relation, as if Janus, the mythical double-faced god, the gatekeeper of past and future, of beginnings and endings, was forever being pulled up by the hair, elevated by a vertical force higher than its two opposing horizontal aspects.

In the doctrine, the Absolute, that which is free from all attendant conditions, that about which no more questions can be or need be asked, is the perfect unity or identity between opposing realities – subjectivity/objectivity, finitude/infinitude, necessity/freedom, unity/plurality, form/content, theoretical/practical and so on. The Absolute is the consummate purity of a universal concept or notion that manifests itself through, but also out of, the manifold of phenomenal reality. The Absolute is the *ideal* of idealism – not only that which is perfectly purified through the ideal realm, but also that which *is* the ideal realm, in all its perfection. This absolute idealism requires something more than an operation of cognitive understanding and knowledge, more than that which Enlightenment rationality can supply. It requires a doctrinal move grounded upon the belief that such an ideal is possible and made present in the here and now. The Word becomes flesh and dwells among us; the kingdom of *Geist* is at hand.

Or so the tradition has understood. And thus, whether in its left or right manifestations, whether the dialectical process becomes wholly historical and the Absolute wholly materialised through a revolutionary impetus, or whether the dialectical process becomes part of a metaphysical or God-driven process and the Absolute a confirmation of rational metaphysics or Christian (Protestant) orthodoxy, there is no return. The revolution, despite its etymology, must carry out the overcoming of existing structures towards the future of a new realm that shall never look back.² In its contemporary sense, if one is still looking back to Kojève, the Absolute God-spirit becomes the religio-political apotheosis of a conservative and neoconservative agenda that sees history in a steady march towards its 'end', an eschaton in which liberal democracy, and all its attending structures, from the free market to the Church, unfettered (oxymoronically) both by and from the State, will usher in a utopian condition free from ideological conflict and division.³

And yet – Hegel returns. And he returns anew, outside this tradition of eschatology and progressivism. The inherent paradox in returning anew is precisely what the following pages will explore. What does it mean to properly return in a manner that is more than mere modification, in a manner wholly and utterly new? Whatever it means, it is neither system nor doctrine. We have to conjure up something different. Or Hegel does. Or we and Hegel do *in concordia*, and in the name not of a Hegelian*ism*, but of a Hegelian appropriation, and one akin to how Ricoeur understood the term 'appropriation' as a hermeneutical sublation, though now recalibrated as a re-appropriation.⁴ Reading again Hegel, anew. Reading again a new Hegel.

But here arises the difficulty – and we can state this right at the outset – a difficulty that plagues all of Hegel's own work. For in reading again Hegel, in re-appropriating Hegel, as the Hegel who remakes himself, as the Hegel who moves on from or outdoes himself, as the Hegel who – and here is the rub – *negates* himself, how can we *come back* to Hegel?⁵ How can we revisit him, re-read him, take him up again, *as Hegel*? For our very re-appropriation cancels him out, even if it claims to preserve him. What does it mean for Hegel to return to us, with us, even within us, when what returns is no longer Hegel proper? How do we appropriate positively a negated (and negating) Hegel? How do we let the negation of Hegel come alive, remain alive? What levels of trickery do we need here? What alchemical process, what spontaneous combustion?

System and doctrine, however construed, are of little use at this point. What will be required instead is something of an invention. An *art*. We need to find the art of re-reading anew, of re-appropriating an ongoing paradox. And we find it, precisely, with-in Hegel.

RETURNING BEYOND BINARIES

Let us therefore begin again.

The past fifty years have seen a resurgence of Hegel, both within Hegelian studies under the banner of philosophy, and within numerous other disciplines and interdisciplines. At first glance, one might argue that it is politics that drives this resurgence, either from the right, as in Francis Fukuyama's earlier appropriation of Kojève's reading of Hegel, which in turn was appropriated for a certain neoconservative agenda,[6] or from the left, as in those who still hold to some Marxist version of a political and economic critique (Badiou, Žižek, et al.). But religion increasingly figures in these appropriations, as it does in the very extreme politics we face each day, whether from a religiously driven political position, or from a specifically theological frame, running the spectrum from Peter Hodgson's embracing of Hegel's philosophy of religion for mainstream Christian theology,[7] to the radical works of Thomas Altizer, which continue to employ Hegel as the grounds of a 'death of God' theology.[8]

Within philosophy, the resurgence has also been on two sides: the Anglo-American axis of analytical philosophy, and the Continental axis of largely French and German, though now increasingly Italian, philosophy. We will deal with the latter axis more fully below. Frederick Beiser, a Hegelian in the former axis, has recently puzzled over this resurgence.[9] In the Anglophone domain, he marks the 1960s as a turning point, with a renewed cultural and intellectual interest in Marx and Marxism. But when Marxism recedes, and as a viable economic and political system in the West finds its tombstone amongst the rubble of the Berlin Wall, he marvels at the lingering, even growing, interest that carries through to the new millennium. For Beiser there are two camps of interest: those who keep their focus on the metaphysical – Charles Taylor's elaboration in the mid-seventies, seminally[10] – and those who keep their focus on the non-metaphysical.[11] In the first, the metaphysics of Spirit coming to self-consciousness and self-manifestation takes precedence. In the second, the postulation of categorical and normative systems within a more positivistic context takes precedence – ethical, political and social realities benefit from Hegelian theories disenchanted of the Ariel Hegel had named *Geist*. Beiser sees strengths and deficiencies in either camp, and ultimately wishes to see both give way to a kind of synthesis, largely in the form of a historical-critical approach that had once been the *modus operandi* of biblical studies: a return to Hegel's *Sitz im Leben* would uncover how the metaphysics of his day – concerning, say, the philosophy of nature – might better inform us today about how to read Hegel *qua* Hegel, and his tremendous influence over the last two centuries.

The division between metaphysical and non-metaphysical readings of Hegel creates an impasse that is, at the very least, 'puzzling'. How do we read the two approaches *together*? Beiser's response is decidedly to move to the non-metaphysical side – historical-critical research. And his suggested areas that remain under-examined in this regard are ones we might expect from such a position: *Naturphilosophie* and natural law within the context of social and political theory.[12] But to place the perplexing problem within the binary of the metaphysical/non-metaphysical is not dissimilar to what Hegel challenged in his own reading of his contemporaries in his 1802 *Glauben und Wissen* (*Faith and Knowledge*). In order to re-read Hegel anew, then, one needs to go back to the early Hegel, and recall what he and his contemporaries were trying to overcome.

In the post-Kantian world of the late eighteenth century, the polarities inherited from the Cartesian move – subject/object, self/other, mind/body, nature/freedom and so on – had yet to be resolved. At first excited by the Kantian attempt, which made mind a fundamental participant in the constitutive nature of the objective world around us, Hegel soon came to see that little progress had in fact been made in overcoming subjectivism. By bracketing out the objective world *in itself*, and dividing reality between the phenomenal and the noumenal in such a way that 'reality' could only ever be conceived in terms of the phenomenal, and never the noumenal, Kant's idealism, though an advance, had far to go. Subjectivism was still the sovereign lord, while objective realities in themselves were not merely serfs in the kingdom of the internal but banished altogether. It is precisely this banishment that, for Hegel, kept post-Kantian thinkers such as Jacobi and Fichte bound to the irresolutions of the Enlightenment, and kept *Glauben* from ever rejoining *Wissen* in any unified manner. Even in these early stages, before the *Phenomenology of Spirit* had been fully conceived, Hegel was working on the 'puzzle': how to read nature (*phusis*) and the metaphysical (*metaphusis*) together; how to dethrone subjectivism and bring objective reality back from exile, without either killing the king or exalting the peasants as royalty themselves.

The Hegelian approach, then, was always, at its very inauguration, an attempt to overcome the binaries of modernity, without wholly sacrificing the modern spirit of self-consciousness born from the *cogito*. That Hegel had doubts about the ability of the previous 'philosophies of reflection' to unify the two sides is unmistakable; but he never felt consciousness could be bypassed, and by the time of the Jena period and the *Phenomenology*, as we know, the unfolding of self-consciousness *across the divide* was taking place. Thus to continue to read Hegel from either a metaphysical or a non-metaphysical position is not to re-read Hegel, or even Hegelian thought, but to re-read the Hegelianisms from which, uncontestably, so much of our last two hundred years of Western intellectual and socio-political history has been formed, and which can now only progress towards their respective ends: for the left, a purely materialist acquiescence to global liberal capitalism, and for the right, a transcendental triumphalist affirmation that, for all intents and purposes, global liberal capitalism is a *fait accompli*.

But to re-read the Hegel that went beyond the binaries, we must situate neither ourselves nor Hegel in the metaphysical or the non-metaphysical. We must instead conjure up a third position, which is neither of the two, but which is neither *not* of the two. We must invoke Hegel and his sublation, in other words. And thus here at this juncture of late modern philosophical history Hegel returns, and yields a characteristically Hegelian third term: the post-metaphysical.

Now the prefix 'post-' continues to produce an eczematic reaction for many a scholar. Especially to those within traditional disciplines of Anglophone thought, it has come to signify much of the worst excesses in recent scholarship, beginning with the master term itself, 'postmodernity'. And yet no one can properly understand the utilisation of this prefix, in any of its manifestations, without a clear notion of Hegel's *Aufhebung* (sublation), from which principally these manifestations gain their meaning and their thrust. 'Postmodernity', we know, does not refer to a state or condition beyond a modernity left irretrievably behind. But neither does it refer to a mere prolongation of modernity, which remains the same at its core, even if undergoing a constant change of attire. Postmodernity is both a cancellation and a preservation of modernity, both an abrogation and retention of its most salient features. We might say, without yet resorting to the language of synthesis, that it is a problematising: an acknowledgement that the salient features are precisely salient for the wrong reasons, or, perhaps better, that they are salient because what they propose or they claim they cannot deliver. (This is precisely what Hegel claimed of the philosophy of reflection in his day.) If we have now left postmodernity behind it is because we accept that the proposals and claims were always, and only ever going to be, incommensurate with their delivery, and we realise we must therefore turn now to other proposals, other claims. But our turning will remain within the orbit of the modern, if only because we have not yet figured out how to make a proposal or a claim, of any kind, outside the gravitational pull of self-consciousness and its freedom. And this accounts, in part, for why Hegel remains in force.

The post-metaphysical is caught in the same dilemma as that which produced it, postmodernity. It both abrogates and retains the metaphysical. The only difference is that we have yet to advance beyond it, except by means of complete exclusion (Beiser's move, leaping from the metaphysical directly to the non-metaphysical). Those working within the currents of the post-metaphysical, largely from the Continental sphere, are committed to working out its features, by which they mean to problematise the transcendental nature of the metaphysical realm, and its claims to ultimate certainty, without necessarily precluding the possibility of what that nature once meant to entail: 'truth', 'God', 'value', 'right', 'the Good', 'the Absolute' and so on. That so many of these thinkers have turned, or returned, to Hegel, suggests to us that Hegel himself was always, from *Glauben und Wissen* onwards, operating in a post-metaphysical space, even if there was not yet the philosophical and cultural ripeness to name it such.

Or we could say: to think post-metaphysically is not possible without Hegel, or not possible without going back to Hegel and reading him anew. And this is what remains so intriguing about Hegel: to go beyond the contending Hegelianisms of the previous two centuries, to get beyond the mutually exclusive categories of the metaphysical and the non-metaphysical, one needs to go beyond Hegel by going back to Hegel. That is to say, negotiating the paradox of 'returning anew' is only possible through some appropriation of Hegel. To go back to modernity *anew*, to go back to metaphysics *anew*, to surpass these conditions or these realms by staying within them, requires Hegel. Why? Because of the *Aufhebung*? Because sublation is the prefiguration of the 'post-' condition? No. At least not in the first instance. It is because any sublation is predicated itself upon an originary force or potency: *negation*. Negation is what allows us to return anew. And Hegel was the first philosopher within the context of modernity to understand this fully and comprehensively, and to make it the crux of his thinking.

ON NEGATIVE TERMS

Negation has always received bad press. Understandably: it eradicates, it takes away. And not just the dead leaves, but the roots. Moreover, as human action, it is perceived to operate by way of misanthropy, or philistinism. It thus cousins nihilism. And not just any nihilism – for there are many – but the nihilism that is wilful, that is intent on destroying all forms of stability and order, and placing humankind, or reality as such, into the vexation that is nothingness.

But we need to be wary of this 'negative' perception, for certainly Hegel never saw it in such terms: a negativity that reduces to nihilism. This is what we might call a 'bad negative', following Hegel's 'bad infinity'. Negation for Hegel is not simply an absence (of existence, of meaning, of purpose, etc.). Negation, rather, is always active. And its activity drives things forward. It is a *movement* and a *force*. And forces, we know from Aristotle, cannot be static absences, or lead to static absences. If negation becomes wilful or intentional, it is because negation has a motive, in the sense that it motivates, it moves and makes motion, it supplies the energy by which some form of kinesis takes place. And thus it is originary, providing the origins for something to come to be, a becoming, even if a becoming by means of a nought.

And so, here at the outset, let us also be wary of the terminology. Negation, negativity, the negative, nought, nullity, nothing, nihilation, nihilism – the range of related terms may seem abundant, even in the English language alone, but the difference between each word is not always straightforward. Nor is their translation, either from one language to the next, or from one context to the next. We find, instead, that of all sets of words, these that revolve around nothing lend themselves most to highly personal appropriation, to eccentric use, to semantic idiosyncrasies, to neologistic alteration. What is nothing for one person becomes a very different nothing for another. There is a certain plasticity, then, that characterises nothing and all its family resemblances.[13] So in this spirit – the spirit of a pliable term that at the same time applies its own force – and before we proceed too far, let us attempt to put down some of our own markers.

'Nothing', as Hegel's *Nichts*, we might say is the indeterminate or predeterminate condition of a void, the absolute state of an abyss, if by 'absolute' here we mean completely free from any dependency or admixture, wholly and undifferentiated nothingness in all its vacuity. If it has any comparison, it is the purest of oppositions to existence or being, but so pure it does not, as the ancient Parmenides taught, lend itself to comparison, and therefore to thinking or to articulation. At least, not to easy, ready-to-hand thought and articulation, even that of an 'opposition'. In fact, we quickly see that language begins to show its limitations, since all of the descriptive terms we may marshal into use already infiltrate the absolute purity of nothing with some other positive assumption: comparison implies distinction, opposition implies differentiation, indeterminacy implies determinacy, mixture implies distinguishable elements, void implies a container or containment, abyss implies depth, and so on. Worse, we must place all these terms in some kind of propositional structure even to approximate description. Even the interrogative of nothing frustrates us. What *is* nothing? But this question makes no sense, when the verb 'to be' stands as its fundamental opposition. It is difficult for us to contemplate nothing, therefore, much less to speak of it. Thought and language keep vanishing into the fissures of their own metaphors.

For some, the way forward is by way of an apophaticism, to speak by way of negative attributes. Hegel may appear to adopt this approach himself: nothing (*Nichts*) can be spoken of as non-being (*Nichtsein*), in opposition to being (*Sein*), for example. But this too has limitations, for this presumes that nothing and the negative can be reduced to an attribute, and Hegel never wants to suggest that nothing and the negative are mere attributes of something fundamentally positive. The nothing functions *on its own terms*. But what can this mean, and what is the 'negative' that characterises those terms? The negative (Hegel's *Negative*) begins to take into account some sense of opposition found when nothing comes up against something other. The negative is that whereby the operation of nihilating activity has taken or continues to take place, whether through counteracting, through emptying, or through nullifying either the positivity of something or, more significantly and more radically, its own negativity. And 'negativity' in turn (Hegel's *Negativität*) is the condition that allows that operation as a possibility, and that in fact realises that possibility, even if paradoxically.[14] All these notions, to be given fuller elaboration below, are, we can see here now, interconnected.

Nihilism, for its part, is a general conception, and one Hegel rarely solicits, that either assumes the *state* of nothing or the *nihil* as reality, or desires to achieve that state. (But the assumption and the desire can take many forms.) We will see further below why Hegel does not admit negativity or negation as a 'state' per se, much less as an 'ism'.[15]

But where, amid all such definitions, are we to place the term 'negation', the term we have adopted for our title? As appropriated, negation perhaps carries the most variation in meaning. Negation (Hegel's *Negation*) is often described as part of the dialectical process, and as such it presupposes something upon which its force can be directed. It becomes the singular activity of opposition by which something positive is negated or made into nothing, which is then further negated to create something yet again, the third term in the synthesising schema. As an activity it therefore comes after the realities of being and nothing to mediate them. And this would bear out well within the regime of the standard dialectical structure: an antithesis already has built into it the thesis that it opposes. Such structure or schematism suggests two things.

First, it gives primacy to negation as a mediating role. Negation comes after both being and nothing in the function of agency, and thus relies on the dialectic for its potency. This is how Judith Butler, for one, first understood it: 'a principle of absolute mediation', or, 'a differentiating relation that mediates the terms that initially counter each other', and therefore is a 'final realization of self-consciousness'.[16] In such a role, negation is always further down the dialectical process, and culminates in a 'final realization' that its movement necessarily entails.

Second, it gives primacy to positivity. The act requires something opposite against which to act, a posited object that stands in the way, and negation therefore becomes reliant upon that object for its effective enactment. If this is the case, then negation may be a power that can effect change, and maybe even towards the new, but it loses its ability to 'return anew' in the sense of a purely generative potency operating prior to the subject/object split. And this loss jeopardises the very change and newness it hopes to effect and sustain. We have seen this loss worked out in more historical terms, where, bound to a material dialectic that has marked the Hegelianisms of the left especially, negation becomes a revolutionary reflex with questionable power. All our revolutions of the left, from Marx onwards, have seemed destined to give way to positivity. The lesson from modern history, a lesson Hegel himself teaches in relation to the French Revolution, is that revolution is followed by the *restriction* of the motions of turning or revolving, and often through terror – if not terror in the form of bloodshed, then terror in the form of suppression or manipulation. And here we might invoke the young Hegel's prescient phrase, and say that to remain in the spirit of modern revolutions is 'to stay immovably impaled on the stake of the absolute antithesis'.[17]

But what if negation did not presuppose positivity, or only positivity? What if, radically, the active force of negation could also be directed at negation itself, the force countering its own force in the very purity of its action? With this possibility, the matter no longer concerns an antithesis opposing a pre-given thesis, but rather concerns the force itself, divested we might say of an object upon which to act, whether positive *or* negative, an isolated primordial force simply in act, even if against its very own self (but now what would 'against' mean here, or what would 'for' mean?). Hegel, we will see later, provides us with a term for such force that is neither/nor and both/and – 'becoming'.[18]

To take negation beyond its dialectical mediating role, beyond simply the agency that brings about synthesis, is to follow Angelica Nuzzo's argument that Hegel's nothing is best seen as the arch-originator of all negativity in Hegel, an *Ursprung* of creative power that 'begins' pre-dialectically, not so much in the chronological sense of coming before, but in the originating sense of furnishing the very possibility for dialectic's own movement to unfold as negating activity, and be maintained in its rhythmic oscillation (negation, negation of negation, negation of negation of negation, etc.).[19] Our sense of negation here would be homologous to Nuzzo's nothing, except as nega*tion*, as we'll see directly, it would already carry its own embodiment with it. This is its great potency, its great creative power. And it is by means of this power that we are able to 'return anew'.

The homology just suggested between nothing and negation extends itself further. What we hope to show in the course of our discussion is that, by retaining negation as an active, actualising power of a creative movement, the subtle distinctions between all the terms above in this family resemblance of negation begin to break down. The primordial sense which we might give to indeterminate nothing fuses with the negative that might arise only by means of dialectical *result*; the condition of negativity that grants and realises the possibility of negating opposition becomes no different to the mediating role of a negation that follows upon any previous condition or sets of conditions and that allows them their dialectical sublation. The term *negation*, by virtue of its ongoing activation, becomes then the prioritised term for the common activity in all these cognate terms, and more importantly, for the very action that works *on itself*. Let us look at this self-activation more intently.

THE ACT OF NEGATION

The Latinate suffix '-ion' denotes a verbal action made into a noun, or made, we might say, objective, perhaps even concrete. The act of negating, then, finds its object or concretion in *negation*, yet not in any static way, but in a manner by which the action continues even in its nominalisation.[20] Negation then becomes the *actual* or *actualised* activity of negating, of making void, of emptying, or of nihilating. Negation actualises by doing. Negation *does*. This is more proper than to say 'negation *is*'. We might compromise and say negation is always in act, but the 'is' here belies a certain predilection towards something essential(ised). Negation does not have an essence as such, any more than it has an existence. Negation acts, wholly and without remainder. Negation actualises its own verbal motivation. And in this perpetual actualisation it is more than just a mediating force in a dialectical arrangement.

But what negation actualises is first and foremost itself: negating is made negation, or negating makes negation. Negation *does*, yes, but negation does first *itself* – by doing itself over. What marks negation – the verb turned noun – is its capability of operating internally, upon itself, in the circularity of a negation of negation. This is not to be thought simply as some form of litotes, which restores positivity by way of the negation of its opposite. It is not the negation of the negative, bringing us now back into something positive. It is rather the negation of negation, which is to be thought of as the force acting wholly within, so that the binary of the acting subject upon an object is forever internally destabilised, and the subject/object division necessarily breaks down. What is being negated in the negation of negation? The force that negates, but that also must remain in order to negate. Hegel's predecessor J.G. Fichte coined a German term to capture such an auto-condition: *Tathandlung*, often translated as 'fact/act' (or just as easily 'act/fact'). What is taken as a given is something *self-produced* as that given. For Fichte, the 'I' as self is both a fact and the act of establishing that fact, both positing and being posited (by itself).[21] Translated to negation, which Fichte had already suggested to Hegel (the self and the not-self interrelate by means of a synthesis),[22] the negating act acts out its own factuality, or it factors its own actuality. The *Tathandlung* of negation might therefore provide some comprehension to the apparent oxymoron 'there is negation'. Negation is neither strictly a fixed noun nor an existing 'thing' but rather, to rephrase Fichte, its own act of self-opposing. In this act, the positive/negative division begins to breaks down. What we are left with then is nothing, and yet everything. Neither, but also both. The same as not the same. Difference conflated with identity. A mediation in its immediacy, or mediation in immediacy.

Now in this act, we have also said that negation is originary, is generative.

What ultimately do we mean by this? As Aristotle sits behind the notion of act (and the act of notion), so too he sits behind this act as generative. Motive power, as the power to move, and the corresponding movement that it produces, were for Aristotle part of the essential features of what he called, in the *Metaphysics*, 'a principle and primary being'.[23] The power behind movement might stem from a material source, as when the wind moves a branch, or it might stem from an intellectual source, as when we decide to move a body part or make a change of any kind (change being the manifestation of movement). Ultimately, we recall, it stems from an eternal source, which must first enact the power of movement as the most primary and irreducible of sources, and this we know as the Prime Mover. The primacy of the act, for Aristotle, or 'purposive activity', as Hegel calls it in reference to Aristotle,[24] is how we must define the actuality of existence, and it is important to emphasise that the principle function of mind, as intellect (*Nous*), is to *move*.[25]

The notion of the generative ability of the mind to move came back into full force in modernity following upon the philosophy of Kant, though now with the Aristotelian notion of *Nous* giving way to that of consciousness. The Kantian mind, as Reason, was not just participative in reality, but constitutive of reality (at least the reality available to us, phenomena). It was Kant's direct successors who wanted to take the idealism of the Kantian scheme and recover in it the active and practical elements that, through consciousness, or more to the point, through self-consciousness, bring the theory back down to ground, so to speak. They sought, in Michael Vater's words, 'that peculiar activity of self-consciousness where act and awareness fully coincide.'[26] As Fichte famously stated in his *Science of Knowledge*, 'hence the "I am" expresses an Act',[27] echoing, or perhaps influencing, the even more famous lines of his contemporary Goethe, in the First Part of *Faust*, when the main character re-translates the opening of John's Gospel from 'In the beginning was the Word' first to 'In the beginning was the *Mind*', then to 'In the beginning was the *Force*', and finally to 'In the beginning was the *Act*'.[28] The young Schelling, arguably Hegel's most direct and most profound influence, was to write in 1800: 'Self-consciousness is an act, yet by every act something is brought about in us.'[29] And thus:

> The concept of self arises through the act of self-consciousness, and thus *apart* from this act the self is nothing; its whole reality depends solely on this act, and *it is itself nothing other than this act*. Thus the self can only be present *qua act* as such, and is otherwise nothing.[30]

So the idea of movement, or of activating force, returns, ironically, out of an idealism insistent on making mind a generative force not just in the understanding but also in the constitution of reality. And yet this active nature of mind must conjoin us to the practical nature of our existence. As the slightly older Schelling would say in 1809, 'The entire new European philosophy since its beginning (with Descartes) has the common defect that nature is not available for it and that it lacks a living ground.'[31] What was to return it to such a ground was the dynamism or organicism of consciousness and self-consciousness, which for Fichte was connected to the social realities of others around us, for Schelling was the essential ground of all freedom, and for Hegel would famously culminate in a world spirit, *Geist*.

But as much as this thinking of conscious mind as self-conscious movement was gaining ground during this very fertile period in German history, it was Hegel who placed the source of its power most directly in negation. Only he saw that for consciousness to remain free – and freedom was the concern, and indeed the hallmark, of all these post-Kantian thinkers, as much as it was for Kant – in order, that is, for consciousness to remain open to the new, uncurtailed by past dogmatisms, and yet not incarcerated by its own subjectivism, or by a thoroughgoing solipsism, in order, ultimately, for the subject and the object to fully coincide, negation must be the internal driver, even against its own forces.

We might elucidate this matter this way. If self-consciousness becomes the new motive force by which, practically speaking, all reality 'comes into being' for us in modernity, the notion of 'creation' then becomes problematic. Creation, as the creative act, would not only entail a positing of the positive, as all definitions of creation presume, but also, and necessarily, a *self-positing*. But how could this self-positing break out of its own subjectivity if its objects of creation – only ever itself – are always positive? This would entail no movement between spheres of subjectivity and objectivity since both are posited as the same – subjectivity. And some notion of a 'creation of creation' does not help us here, since positivity upon positivity only worsens the case. As Habermas puts it: 'The modern manifestations of the "positive" unmask the principle of subjectivity as one of domination.'[32] Only when some negating notion is introduced can self-consciousness find a way through the bars of its own self-imposed cage.

Why does Hegel return? Because we cannot let go of our modern inheritance: self-consciousness as the *sine qua non* of active freedom. We must, at all cost, retain the individual freedom to act according to our own consciousness, because it is our consciousness that provides the active, motive ability to effect change, while it is that change, as act, that keeps freedom alive, and always open to the new.

But consciousness has had a rough ride since Hegel, who understood it had its 'unhappy' moments, as he famously described it in the *Phenomenology*. And unhappiness was to give way to unmitigated perturbation under the gale force of Nietzschean critique (consciousness as an unsustainable unitary faculty, even an '*accidens*'),[33] of Freudian critique (consciousness as driven by a subterranean force beyond its immediate control), of Lacanian critique (consciousness as constituted by fracture and *méconnaissance*),[34] and of postmodern critique (consciousness as decentred and dissipated). Yet despite all these tempests and storms, the discourse of freedom remains paramount, and life after postmodernity is no less committed to its underlying principle: that we are autonomous beings wholly free, who should remain wholly free, to act out who and what we want to be. As long as we remain beholden to this principle, Hegel will return, for only in Hegel does consciousness, through negation, permit something other than self-positing in freedom.

Concepts of nothing, negativity and negation have increased in the last century precisely because of an acting, motivating, creative consciousness that cannot hold its own. If modernity presents us with the former, the creative, productive consciousness, and postmodernity the latter, the inability of that consciousness to hold its own, then our present condition – call it what we will – presents us with the need to move forward in freedom in the face of the knowledge that this freedom is nothing other than the *object* of our active creativity, rather than its subject. That is, freedom is only something we have invented, and not the very constitution of consciousness. And if this is the case, then negation will inevitably set itself into motion. And Hegel will return. For where does one get the freedom to invent freedom? Only from a negation that first negates the constituting freedom in order to allow the constructed freedom to arise. But how can freedom remain free if it is constructed from something that is not free? In Hegel's language, only when the Slave recognises that the Master is truly not free.

THE ART OF NEGATION

The question of consciousness as the irreducible creative force, creative because it acts upon the world to unify subject and object, and therefore 'constructs' reality as we apprehend it, was being worked out within the Romantic aesthetic of Hegel's time. Even Hegel's closest German contemporaries – Schiller, Hölderlin, Schelling – were convinced that the myths we create, poetically, are the only way back to a unity of the conscious, subjective mind with the objective world it lives in. As Schelling states, 'that which appears to us outside the sphere of consciousness, as real, and that which appears within it, as ideal, or *as the world of art*, are also products of one and the same activity.'³⁵ But Romanticism, for Hegel, was no further forward in breaking free from subjectivism; the Romantic aesthetic generally made the subjective even more exclusive, and thus, as it were, pulled down the blinds on the prison window, while even his German friends and contemporaries, in making the productive consciousness primary and irreducible, preceding the world that is out there, could not avoid the ultimate incarceration that subjectivity would entail. Hegel by no means wanted to abandon the power and role of art. Even in the early years, he felt that art was instrumental in bringing consciousness to a new level. The unificatory powers of consciousness were what he objected to, the positing of positivities in full harmony with each other, and thus also art as positive unification. For this did not unify *absolutely*; it unified only conditionally, based on the conditions of the self as conscious subject. What Hegel needed to find, therefore, was a way out of this dilemma, a way for active, motivating, creative consciousness to free itself of itself. This he found by transferring the powers of productivity and motivation away from consciousness as a positing force and towards negation as a self-separating or self-cancelling force inherent within consciousness. Negation thus becomes not an extrinsic power, imposed from without, but an intrinsic dynamism, always operating from within. And if negation becomes a conscious force, it does so only as conscious of its own *self*-negation, or conscious of its own *as* self-negation. (But this 'own' is also not its own, proper. If it was, it would simply posit itself, positively.) Or, consciousness inheres in the powers not of unification but of self-diremption, the very negation of unification.

What Hegel introduces into our modern consciousness, then, is that which we increasingly struggle to come to terms with in our contemporary world: the *diremptive* nature of our own 'productivity'. Why does Hegel return? Precisely because we cannot yet fathom how it is that, as emancipated selves wholly free to construct ourselves and our world as we see fit, we repeatedly turn productivity into catastrophe. Rather than unifying our inner selves with our outer world in the harmony and freedom of a liberated consciousness (and conscience), we continue to dirempt these two realms – both inter-diremptively and intra-diremptively – in the name of a freedom that contradicts liberation. The supposed unificatory powers continue to wreak havoc through strife, domination, oppression and coercion, whether in political, military, social, ecological, religious or psychological arenas. The last half-century is not only testimony to a deep existential crisis in response to the atrocities of two world wars, and to the Holocaust as the consummation of Western diremption, but also about our inability to stem or resolve global strife thereafter (Korea, Vietnam, the Gulf, Afghanistan, Iraq, etc.), the inability to eradicate genocide as a State option, the inability to breach, never mind dismantle, terrorism, the inability to decelerate poverty, the inability to curtail corporate avarice, the inability to keep ethics apace with technology, the inability to safeguard the environment, the inability to neutralise media power, the inability to grasp the continuing fragmentation of modern art, the inability to ground personal relationships in constancy and perdurability, the inability to comprehend the self-loathing of the solitary figure on the rampage, the inability to provide spiritual contentment, the inability, in the end, to reconcile the fact that our constructs of freedom are ultimately binding, discordant and ruinous. And all this is our own product, the productivity of division or of lack. How, through freedom and mastery, have we become so dirempted, so self-estranged? Hegel returns, both by way of a diagnostic, and by way of a possibility.

And to see the bi-directional nature of this Hegelian path – returning anew – we need to focus on this question of art more directly, murmuring as it has beneath our introductory thoughts so far. Though gains have been made more recently, art continues to be largely under-represented in Hegelian studies. Since Kaminsky's early study *Hegel on Art* (1962),[36] the paucity of full treatments of Hegel's art and aesthetics is conspicuous, despite the fact that Hegel's published lectures on the subject are more voluminous than any other of his topics. The secondary works that do exist either follow Kaminsky and focus exclusively on aesthetics, with little attempt to ground the aesthetics on the intricacies of Hegelian epistemology and ontology (Bungay; Wyss);[37] purposely move away from art's connection to speculative logic, and focus on social realities (James);[38] are edited volumes with varied focus (Steinkraus and Schmitz; Maker; Houlgate);[39] or focus exclusively on Hegel's supposed 'death of art' claim (Lang; Danto).[40] Wicks has emphasised *Hegel's Aesthetic Theory of Judgement*, Speight has elaborated Hegel's employment of literature, Rutter has written about Hegel in relation to the modern arts principally during Hegel's time, while Desmond has been most comprehensive in drawing in all of Hegel's thought to bear on his aesthetics in *Art and the Absolute*.[41] Individual articles have been on the rise, some formatively (e.g. Henrich; Pippin; Houlgate; and Hamacher).[42] But overall the field remains comparatively under-ploughed, as Desmond himself acknowledges,[43] which suggests that many continue to take the Hegelian 'death of art' claim at face value, and apart from a certain historical reading that conforms to 'the system', have deemed Hegel's reflections on art themselves to be moribund.

The 'death of art'. This famous phrase, supposedly made in the Introduction to the *Lectures on Aesthetics*, would seem to define most succinctly the problems we currently face in late modernity: our creative powers of productivity are, 'positively' speaking, dying or dead. But of course this is not what Hegel meant, and not even what Hegel said. If we want exactitude, he said: 'In all these respects, art, considered in its highest vocation, is and remains for us a thing of the past.'[44] Now what are 'these respects'? They can be found in Hegel's statement only a few pages earlier: that true art, as fine art, 'only fulfils its supreme task when it has placed itself in the same sphere as religion and philosophy'.[45] So that once art subsumes itself in religion, and religion in turn subsumes itself in philosophy, art effectively becomes *surpassed*, and what we are left with is a *philosophy of art*. But since the self-conscious spirit is precisely what drives art, and since the self-conscious spirit thinking its own self is precisely what *constitutes* art, then an art that thinks on itself, that judges itself as an idea, that goes beyond the sensuous, material, tactile realities from which it first came to life, is precisely the natural and appropriate result. And this result is *aesthetics*: no longer the motivating creative power in its irreducible form, but reflection on the productivity that that power has already made manifest in the work. Aesthetics, we might say, is the death of the originary power of art.

But aesthetics is already well down the line, even in Hegel's thought. (The *Aesthetics*, we remember, was never part of the formal Hegelian system, never part of the 'Encyclopedia' – it was only a concatenation of lectures and student lecture notes.) What if we were to rethink that phrase, 'a thing of the past', and consider that in order to recover art from aesthetics, and revivify the activating power of *poesis*, in order to clear the ground once again for the productive force that we most associate with the creativity that is art, we must somehow go *back* to art, must return to it anew? This is what the following pages will set out to do, not in the name of aesthetics, even if the *Aesthetics* will be addressed, but in the name of something more originary, a name we have now already named, negation. And in that name, as force, we will negate the 'death of art', negate the negation of art, and suggest instead an art of negation, one that sees 'art' returned (to) anew as still very much alive, and one that does art, the art of the living, if only by way of its paradoxical self-originating death.

The art of negation itself comes alive very much within the post-metaphysical we have already begun to figure. Derrida's *Glas*, as we will see, sets Hegel across a textual gap from the artist Jean Genet, in a move that negates normal conventions of reading in order to incite us to read anew out of the gap, the diremption, that stands between. Jean-Luc Nancy's ongoing work on Hegel clearly understands the importance of the creative throughout Hegel's *oeuvre*. Following Derrida, Nancy writes: 'The "end of art" is always the beginning of its plurality.'[46] In the centre of his book *The Man Without Content*, Giorgio Agamben employs the phrase 'art is a thing of the past' to bring into relief what he calls a certain self-annihilation, or nihilism, or nothingness, in operation throughout the West. And the artist, he says, 'is the man without content, who has no other identity than a perpetual emerging out of the nothingness of expression and no other ground than this incomprehensible station on this side of himself.'[47] With both Nancy and Agamben, the 'death of art' is not considered as an endpoint for thinking about art, as if it was lost to a cultural nihilism irretrievably, but on the contrary as a beginning point for thinking anew about art, about *poesis* and its praxis. In both, we have the question of negation and art not in the restricted manner of a faded or consumed aesthetics, nor as an unproductive nihilism, but as a re-emergence of a productive force on the very site that has remained so obscured in Hegelianism: the creative and effectual power of the Spirit as imaginative, originary act.

What the art of negation announces, then, as Agamben has already hinted, is something that both brings art into being and leaves it behind (in its 'death'), and that the very nature of this movement must be understood from the question of the originary force of art *in the first place*, before it ever reaches systematisation (i.e. aestheticisation). What if we turned from looking at how art and its history fit into Hegel's system, or how the death of art should or should not be conceived, either within Hegel or within the culture that has inherited Hegel, and instead looked at how Hegel fits into art? Or more precisely, what if we took this negationary movement as the source by which we begin conceiving of productivity in any sense, artistic or otherwise? What if we developed an understanding of this force as rooted, productively, in all our later endeavours, whether religious, philosophical, political, economic, even ethical? This is what the post-metaphysics of such thinkers as Nancy, Žižek and Agamben have been shadowing.

So let us dwell on the originary moment of art, *before* it is art, or what might amount to the same thing, in Hegel's terms, *after* it is art. By focussing on this originary moment, we do not simply offer a consummation of what others have already implied, or, in the case of Thomas Altizer, have tried to work out in the name of an apocalypse. What we suggest is something that is both a coming before and a going after, which together infects the great triad of Hegel's own absolute system at its very centre: art, religion and philosophy. If art has, on one end, played itself out as, according to Hegel, something surpassed, then going before art to this originary moment of negation also plays itself out on the religion and philosophy that supersede it, since this originary moment lies at the heart of all cultural manifestation. If art's 'death' or 'end' is merely an undoing of the very structures that allow art to be what it has traditionally been for us in the West, then, systemically, this undoing penetrates into the very soul of religion, both as the art-religion and the revealed religion of the *Phenomenology*, as it does into the very soul of philosophy. Rather than a triadic movement whereby both art and religion give way to the supremacy of a fully consummate or self-conscious philosophy at the apex, all three corners of the triangle – art, religion and philosophy – become 'surpassed'. That is, in speaking of art as a thing of the past, religion and philosophy must, necessarily, undergo their own negation as well. But this negationary movement does not leave nothing. It takes us beyond the categories or limits of art, religion and philosophy and into something new. It reverses Kant in this respect: it breaks open (disrupts, rends, dirempts) limits. And in doing so, we move into a wholly new interdisciplinary territory, in which art, religion and philosophy are radically reconceived, even beyond Hegel.

This book, then, traces out this 'movement' more directly and determinately, and thereby brings us to this interdisciplinary place in a way that might compel us to rethink the relationship between art, religion and philosophy, and between those other concomitant manifestations to which Hegel's system has been so often attached: politics, rights and ethics. It pursues the negations of Hegel to their logical, and translogical, 'conclusion', if by logic we also include the very logic at the heart of Hegel's science (and the very science at the heart of Hegel's logic). The implications are vast, for not only must the 'left' and 'right' designations of Hegelianism fall away (as, *pace* Habermas, for all intents and purposes they have), but the working out of surpassed Hegelianisms, as things of the past, will lead us to a kind of 'transdisciplinarity', negating the disciplines by going beyond them, to a place where action, in its most primal vitality as (negating) movement, must be reckoned as action towards others, or as ethical action. This would be in the spirit of Hegel as the reconciler of subjectivity and objectivity. For even in the *Philosophy of Right*, Hegel understands freedom and the will to be grounded upon this negative movement, since it is through negativity that the 'I' that wills is determined in its determinacy.[48] It is this very determinacy, perhaps now of ethical action, that would have to be fully addressed in the *after* or 'post' of this surpassing space beyond art, religion and philosophy in their traditional formulations. Here is where Hegel, newly returned, might have something to say about the current negations we see going on around us on a global level. Perhaps here is where a new ethics might lie, an ethics that surpasses even the worst of our negations, or even the worst of our ethics. This would mean a productive ethics, even if not a progressive one.

But first, we must turn back to Hegel.

PART I

THE HEGEL OF NEGATION

CHAPTER 1

Negation's Art in Phenomenology of Spirit

The *Phenomenology of Spirit* (*Phänomenologie des Geistes*) remains one of the great preludes to any philosophical system. It is such a *tour de force* that it has become more widely read than the works of the actual system that it supposedly preludes. What gives it this force, this enduring attraction, is not only the audacity with which it claims to overturn all previous philosophy. It is not only the dynamic propulsion that drives the argument forward with unrelenting verve and purpose, but also the way in which this teleology – which finds such celebrated form in the system that is to follow, the dialectic that leads us to the greatest absolutisation since the Neoplatonics – is undermined by its own internal mechanism. Before the system could be solidified into the philosophical bulwark it inevitably became, either in the hands of Hegel or with those Hegelians who came after him, the daring extravagance of the *Phenomenology*, in laying out its nascent terms with such prodigious ferocity, opened itself up to its own slippages and reconfiguration. This, it might be argued, is what makes the greatest of our philosophers: the ability to build a system of thought that can outthink itself, that can operate *in spite of* itself, maybe even against itself. Plato is perhaps our earliest example, but Hegel had a much closer predecessor in Kant. In this chapter, we will look at the way the unfolding of consciousness, as driven by an originary force we have re-figured as art, works to challenge the dialectic movement we have become accustomed to assume for Hegel and Hegelianism, and in fact works to undermine that dialectic in its most absolute form.

Let us begin by focussing upon the narrative dynamic that drives the *Phenomenology*. Many have remarked upon the dramatic nature of this dynamic, where the text is very much like a theatrical unfolding, whose protagonist, through stages, overcomes continual obstacles before it reaches a triumphal peak.¹ Or it is a kind of *Bildungsroman*, whose main character, Consciousness, develops, by means of successive periods of interior growth and maturation, into self-awareness.² Or it is a textual journey, a text of a journey, of an *Erfahrung*: 'this *dialectical* movement which consciousness exercises on itself and which affects both its knowledge and its object, is precisely what is called experience [*Erfahrung*].'³ And as Beiser informs us, 'Hegel's term "*Erfahrung*" is therefore to be taken in its literal meaning: a journey or adventure (*fahren*), which arrives at a result (*er-fahren*).'⁴ Consciousness is on a journey, and we accompany it in its itinerary until it reaches its vaunted destination. As Hegel says in the Preface to the *Phenomenology* ('On Scientific Cognition'): 'It is this coming-to-be of *Science as such* or of *knowledge*, that is described in this *Phenomenology* of Spirit... In order to become genuine knowledge, to beget the element of Science [*Wissenschaft*] which is the pure Notion of Science itself, it must travel a long way and work its passage.'⁵ Or in the closing pages: 'The movement of carrying forward the form of its [Science's] self-knowledge is the labour which it accomplishes as actual History.'⁶ So the very idea of historical narrative itself becomes an unfolding drama, the very stuff of the modern novel,⁷ and what beguiles us in the *Phenomenology* is that its very textual structure is a kind of conceptual dramatic history driven forward by its own internal mechanism.

The text lays claim, that is, to a self-originating power. But we need to be clear about the nature of this power. It is not that the author, as originator, claims pure originality. Hegel never draws attention to himself as author of the *Phenomenology*. In fact, in the opening lines of the Preface, he repudiates such attention. The author and his or her aims, in a philosophical work, are 'not only superfluous' but 'even inappropriate and misleading'.⁸ And it is more than simply a question of a 'universality' of the subject under investigation. The demand for a prefatory explanation of authorial purpose is, Hegel says, predicated upon a certain anatomical notion of historical-critical analysis: we compare the present work with what has come in the past, and measure it by its truth or falsity in relation to a given traditional acceptance of a factual and historical whole, just as a pathologist studies the parts of the body in relation to the whole corpus that was once alive. The metaphor of anatomy is significant here. For the pathologist requires not merely a corpus but a corpse, a dead specimen, and the autopsy proceeds by dissection of parts in relation to what was once, but is no longer, a living whole. Hegel rejects this dead metaphor, and replaces it with a living one. To understand philosophy proper, to understand how any philosophical system relates to itself internally *and* to what has given rise to it, one must see it in terms of a 'progressive unfolding of truth'. One cannot dissect what is alive, separating off parts, while retaining the life within. One must rather join in the ongoing progressive development of its life, where the different stages, in their unfolding, retain the continuity of an active, living vitality.⁹ So Hegel employs the metaphor of a bud exfoliating into blossom, which in turn yields a fruit. Each stage 'supplants' the previous one, and renders each 'mutually incompatible' with the next. 'Yet at the same time,' Hegel writes, 'their fluid nature makes them moments of an organic unity in which they not only do not conflict, but in which each is as necessary as the other; and this mutual necessity alone constitutes the life of the whole.'¹⁰ It is this organic unity that Hegel will claim for his own text; and it is the generative power of this flowering and yielding that he will try to lay bare.

Such an organicism implies extension to the author, to supplant him or her as the sole authorial origin – a fact which Hegel's 'spirit' will eventually manifest, and which we will try to manifest ourselves in the name of different authors below. But it also works across Hegel's text as a self-contained entity, and by 'across' we mean here both throughout the text and beyond or surpassing the text. The text unfolds the various stages of its own development, leaving behind one stage for the next, yet never fully abandoning any stage in the organicism of its movement. But it further unfolds as a stage amid the corpus of Hegel's writings, between the 'early writings' prior to 1806, and the later, more systematic or 'encyclopaedic' writings, beginning with the *Science of Logic* of 1812. And yet this corpus is also, as Hegel must see it, part of an unfolding within the entire history of philosophy, which itself, as 'system', is the unfolding of the human spirit and the world spirit that are, collectively, *Geist*.¹¹ It is therefore a necessarily living corpus, aware of its own pulsating muscle and tissue within the larger body of its being. So when we say the text originates itself, we mean this in the sense of the metaphor of an organic process, whereby either a body comes to life, animated by a unifying self-conscious Spirit, or the bud, blossoming forth, and yielding fruit, produces its own emergence, its coming-to-be, now in a deeply self-conscious manner. And precisely because this very self-consciousness *is its power*, it 'self-generates'. In Hegel's own words, this is 'thought which begets itself'.¹² But because it cannot do this outside a living metaphor, even in its most conceptual understanding – the metaphor that translates identity across two incompatible spheres – we can talk about the generative function of this organic unity in terms of an *art*, whose formal aspects we can now try to unveil.

Before we take this step, however, let us enlarge upon the conflict that marks the metaphor – any metaphor. Just as the botanical metaphor of coming to fruition supplants the anatomical metaphor of analytical dissection, so the metaphor of bearing fruit supplants *within* itself, rendering the bud dead to the blossom, and the blossom dead to the fruit. The metaphoric trope by definition is engendered by 'mutual incompatibility'. Two differing spheres are brought into identity out of their incompatibility. So we could say that consciousness and self-consciousness, and the movement from one to the other – a making manifest of one's own self-cognizance – is simply a metaphor for art, and perhaps justify the co-opting of a term (art) for a sphere where it does not belong (conceptual philosophy). But if we are to understand Hegel at all, we have to understand that there is no co-opting here in any sense. The incompatibility is *organic* to the translation of one sphere into the next; it is precisely what allows the generative movement. And this we will see in terms of negation, the negation of one side by the other, and the emergence of a new resultant, within the *Phenomenology* as a whole, even if the whole becomes dis-unified as a result.[13]

ART, PHENOMENOLOGICALLY

What is the phenomenon we traditionally call art? Hegel has a very clear answer, though it comes rather late in the movement of the *Phenomenology*. Art, as a creative cultural expression, figures only in the final sections on Religion, which, we know, is well along the journey of consciousness toward self-consciousness, and from self-consciousness as Reason toward Spirit as Absolute Knowledge. References to certain works of art are occasionally made explicit or implicit – most notably Sophocles' *Antigone* in the middle section on ethics and morality. But the general idea of art only emerges properly when Religion moves beyond its natural incarnation, as Spirit that only 'knows itself as its object in a natural or immediate shape', to Spirit that 'knows itself in the shape of a *superseded* natural existence, or of the self.'[14] At this point art in its formal sense makes its entrance, not as *mimesis*, as it might have been in the Aristotelian tradition that held sway for much of Western Europe's history prior to modernity, nor as a function of beauty, indifferent or sublime, as it might have been for Kant or Schiller, nor as an expression, in whole or in part, of the unconscious, as it might have been for Schelling, but as a creative self-conscious development of Spirit moving from outer expression to inner expression.

And yet this formal entrance is not yet a formal expression of *itself* – art per se. It is rather, and remarkably, a formal expression of *religion*. When art figures itself, it does so *as* religion, the 'Religion of Art', or perhaps better, religion in the form of art. This equation of the two realms, religion and art, never again fully appears in Hegel's subsequent work, whether in the lectures on religion of 1827, or in the even later accumulated lectures on aesthetics.[15] In these latter works, the now famous triad of art–religion–philosophy, as a movement of succession or progression, had well solidified. But in the earlier work, though the triadic movement is present in its general contour – absolute knowledge succeeds revealed religion, which succeeds the spiritual work of art – art does not yet have its own categorical function. It is fused together with, only in the end to be subsumed under, the function of religion. What might this tell us about creativity as Hegel first conceived it in the exuberance of the 1807 text, only later to give way to the systemised structure of aesthetics, and become a realm necessarily transcended by religion?

The way into this question lies in the phrase 'thought which begets itself'. If, as we have seen, consciousness is a creative force that generates into being, its manifest generation, in its advanced stages as Spirit that is fully conscious of itself, takes on a particular *shape (Gestalt)*. Religion is that *shape* in which Spirit is fully *self-conscious*, and in which that self-consciousness is made identical both with its existence and with its essence (§680). To attain this unity of being and pure reflexive awareness of being, of Hegel's *in-itself* and *for-itself*, religion must know itself first immediately or substantively, in Natural Religion *(natürliche Religion)*, and then subjectively, in the Religion of Art *(künstliche Religion)*. Hegel writes: 'for the shape raises itself to the form of the self through the creative activity [*Hervorbringen*] of consciousness whereby this beholds in its object its act or the self' (§683). Here the self emerges as the expressed shape and becomes conscious of itself through art, and this *self*-consciousness, and by extension, this *art*, are both raised to the level of religion. Art is thus religion knowing itself as the shaping or active generation of its consciousness as self.

The idea of shaping is significant here. And Hegel owes much to Schiller on this score, more than is often noted. Schiller, in his *On the Aesthetic Education of Man*, had laid out two principle (pre-Freudian) impulses or drives (*Triebe*) that help to unite our subjective experiences with the objective realities around us. The first he calls the sensuous impulse (*sinnlicher Trieb*), which situates us materially, through the senses, in time and finite existence. The second he calls the formal impulse *(Formtrieb)*, which situates us absolutely, through our rational nature, in freedom and infinite existence. At the beginning of the Thirteenth Letter, Schiller then poses the question that has bedevilled modernity since its Cartesian inception: how can these seemingly opposite drives be reconciled with each other?[16] He then offers a solution that will seem all too Hegelian, until we remember that Hegel was only a young post-seminarian, working as a private tutor for a family in Berne, when Schiller's publication first appeared in 1795. The solution is the introduction of a third term, one Schiller calls the *play impulse* (*Spieltrieb*), which 'would aim at the extinction [*aufzuheben*] of time *in time* and the reconciliation of becoming with absolute being, of variation with identity.'[17] If the first impulse encompasses all that we might understand as *life* (*Leben* – 'all material being and all that is immediately present in the senses'), and the second encompasses all that we might understand as *shape* (*Gestalt* – 'all formal qualities of things and all their relations to the intellectual faculties'), then the third is what Schiller calls a *living shape* (*lebende Gestalt* – 'all aesthetic qualities of phenomena and – in a word – what we call *Beauty* [*Schönheit*] in the widest sense of the term').[18] This concept of a living shape returns us to (or more properly anticipates) Hegel's living metaphor above, in which an organicism imbues two incompatible sides toward reconciliation. This living shape, as shaping, or as Art in its originating sense of both bringing to life and bringing to form, becomes for Schiller the task of aesthetic education, a reconciliation of our two impulses in an aesthetic third, something he later describes in a phrase that approximates a synthesis – 'filled infinity' (*erfüllte Unendlichkeit*).[19]

We can now begin to see why Hegel, under Schiller's influence, does not separate out Art from Religion in his early stages of thinking. For if Spirit knows itself in a shape (*Gestalt*, i.e. Schiller's formal impulse) that has superseded natural existence (i.e. Schiller's sensuous impulse or life), it is precisely this shape that, we remember, 'raises itself to the form of the self through the creative activity of consciousness' (i.e. through Schiller's play impulse, or living shape). And this takes place first in Religion, the living shape through which Spirit becomes fully self-conscious. Thus it is that at the beginning of the section on Religion in the *Phenomenology*, 'shape' (*Gestalt*) is introduced as a central concept. For consciousness has not yet appeared in any shape that can be known, or that can allow it to be known, as pure self-consciousness. But Religion now provides that shape, because only in Religion does the shape become 'perfectly transparent to itself', so that consciousness is posited as self-consciousness, and consciousness can now *know* itself as self-consciousness.[20] This ability of self-knowing is precisely the 'thought which begets itself', shaping that self-generates.

We are thus now in a position to see that, under the dynamic of such self-begetting, Religion itself goes through three stages. The first, Natural Religion, shapes things in the form of substance: God/gods as light, plant, animal or stone. Even this has a threefold movement: the first, light, is the shape of 'shapelessness', all-pervading light; the second, plant and animal, is the shape of organic life in its multiplicity (passive/innocent flower religions versus aggressive/guilty animal religions); the third, stone, is the shape of objects shaped by an artificer (pyramids, obelisks, statues, idols, etc.).[21] It is this last, of course, which advances the Spirit, for the creative activity of the artificer, shaping the object of nature, constitutes self-consciousness by virtue of its interiorisation of the exterior, whereby the object of consciousness unifies with the shape of self-consciousness: 'an inner that utters or expresses itself out of itself and in its own self; *into thought which begets itself*, which preserves its shape in harmony with itself and is a lucid, intelligible existence. Spirit is *Artist*.'[22]

The second stage, Art Religion, thus comes into its own at this point. Here, 'Spirit has raised the shape [*Gestalt*] in which it is present to its own consciousness into a form of consciousness itself and it produces such a shape for itself.'[23] Because, then, this shape has gained the form of self-conscious activity; because, that is, it shapes its own shape, religion must be seen as manifesting itself as a form of art – first abstractly, as it moves away from the substances of Natural Religion and into what Hegel calls *'pure activity'*;[24] then as a living work of art, as it moves away from the individuality of the artificer and toward a permeation of community, or of what Hegel calls the Cultus, wherein the stone statue of Natural Religion not only comes to life, but takes on a national or a collective character; and then as a spiritual work of art, as it moves from the generalities of the national form toward a universal 'shape' by means of language ('Gestalt' now in the sense of a whole which exceeds the mere aggregation of its parts, living speech that exceeds each individual speaker or each national tongue).[25] So that finally, in language, religious shaping begins to correlate to poetic forms: the Epic, Tragedy, and then Comedy.

In all these triadic movements, or triads within triads, we can see at work Schiller's movement from the sensuous to the formal to the aesthetically playful and living shape. Art lives as living being, like the stone statue that comes to life at the end of Shakespeare's *The Winter's Tale*: the tragic that is rescued by the comic turn, the 'romance' that mingles the hardness and harshness of tragic fate with the pliable and luminous joy of comic resolution. The statue's form becomes real being: 'The fixture of her eye has motion in't, / As we are mocked with art,' writes Shakespeare.[26] Or in Hegel's terms: 'Through the religion of Art, Spirit has advanced from the form of *Substance* to assume that of the *Subject*', and it makes this advance because, like Shakespeare's Hermione, 'it produces its [own outer] shape' as living subject in self-consciousness.[27] As Hegel states a few lines earlier, 'the actual self of the actor coincides with what he impersonates', just as Hermione coincides with her impersonated statue.[28] This is Schiller's playing at Beauty (our being 'mocked with art'), or his playing *that is* Beauty. 'Beauty... is at the same time a *state of our personality*... at once our state and our act.'[29] Hegel transmutes this language of play into the ethical Spirit of religion, but in the *shape*, and in particular in the shape that shapes itself, or the thought that begets itself, the play impulse and religious formation converge. Shakespeare's play captures this poetic moment of convergence when, just as the statue of Hermione comes to life and steps down, her daughter says to her father: 'her actions shall be holy as / You hear my spell is lawful'.[30]

What we have seen so far, then, is a certain fundamental shaping whose activity, in its purity, does not allow a strict differentiation between generation on the one hand (bringing into being), and religious formation on the other (self-awareness of such generation). In fact, Religion at this stage of the *Phenomenology* amounts to this very creation of self-awareness: the awareness that *poesis* is the generation of awareness itself. Consciousness 'beholds in its object its own act',[31] or 'is a *whole* only together with its genesis'.[32] So ancient Greece, and its poetic forms, become the paragon of this stage, since in Athens, and on the stages of Athens especially (within the Dionysian festivals), art and religion became one, just as, according to Hegel, actor and character became one, and character and audience became one – the inner and the outer in self-identity. The sourcing of this unity confounds the later Hegelian dialectic whereby, in terms of historical development, art must *precede* religion, and religion, in turn, must *precede* philosophy. That is, in its generative form, Art Religion maintains a disruptive possibility, insofar as it brings us to the generative process, the shaping, that stands behind *all* activity, not just art and religion. For this reason, Hegel separated the two in his later thought, and sought to describe that generative process otherwise than an 'art' per se.

And of course Art Religion, even here, stands as the middle term between two other stages, Natural Religion and Revealed Religion: Art Religion is by no means an endpoint. The narrative development continues in the third stage, Revealed Religion, wherein the immediate or substantive side of Natural Religion merges with the subjective side of Religion as Art. If the first is being as substance *in-itself*, and the second is being as subject *for-itself*, then the third is their unity in *'being-in-and-for-itself'* (§683). And this is Religion as having now been revealed, where 'substance is *in itself* self-consciousness' and self-consciousness is in itself substance (§755).[33] This merger as self-conscious Being, now in its capitalised or absolute shape, is the revealed God of religion, who knows itself in the immediacy of its self as Spirit. And when it takes this one stage further, when it knows this immediacy as self-consciousness, or, when its actual self-consciousness is the object of its consciousness (§788), then it knows itself absolutely, and religion gives way to Absolute Knowing, in an apotheosis of meta-consciousness. This is the fulfilment of the great narrative movement within the *Phenomenology*. Thereafter, that movement is rendered with more constituent phases: it begins with art, then moves to religion, and finally consummates in philosophy.

But before we let Revealed Religion carry us away from the heights of Mount Parnassus, to an absolutised sphere of idealist philosophy, and the dogmatic Hegelianisms that have so often reached into the sky from the mountain's peak, let us freeze again this moment when art and religion are self-identical, when unity is shaped by 'thought which begets itself' as the purity of the creative act. And let us ask even further: what exactly is this generative force that drives the act or the activity of creation in the first instance? What fundamentally drives Spirit toward religion in the form of art? Or, more precisely, and with Schiller still in mind, what exactly *drives* consciousness itself? In our post-Freudian world, we have become well adjusted to this question, and well accustomed to the answer. But for the pre-Freudian Hegel, the question is more acute: what allows for the *becoming* of consciousness as a shaping self? Here we need to leave Religion and Art Religion, and return to earlier and more nascent formations within the text's narrative.

THE SHAPING OF NEGATION

In the Introduction of the *Phenomenology*, Hegel writes of the necessity of the movement by which consciousness moves to a new object, and in this connection he speaks of *origination (Entstehung)*. Yet he says that we do not have an understanding of how this *origination* takes place: 'But it is just this necessity itself, or the *origination* [*Entstehung*] of the new object, that presents itself to consciousness without its understanding how this happens, which proceeds for us, as it were, behind the back of consciousness'.[34] So where in consciousness exactly can this generative force itself be found or located? Where is the 'back of consciousness'? Though the answer is best worked out in the following *Science of Logic* of 1812, as we will see in the next chapter, here in the *Phenomenology* we get a preliminary version. Its shape is not fully formed, but it nevertheless emerges, at least spectrally, and under shifting guises, as *negation*.

Here too, Schiller has had his influence. In his *On Aesthetic Education*, Schiller describes Beauty, we saw, as the linking together of two opposing conditions: life as substance (matter, passivity, sensation), and shape as thought (form, activity, thinking). Beauty is the 'intermediate condition'. But in the Eighteenth Letter it is arrived at by a certain synthesis of the two sides: 'Beauty *combines* those two opposite conditions, and thus removes the opposition.' We quickly see that this synthesis is a sublation (or *Aufhebung*), for Schiller immediately adds: 'But since both conditions remain eternally opposed to one another, they can only be combined by cancellation [*aufgehoben werden*].'[35] It is likely that Hegel drew his notion of *aufheben* from Schiller here, where Beauty must not only destroy but preserve both sides of the opposition. This is its reconciling function. In the subsequent letter, Schiller elaborates: 'So we arrive at reality only through limitation, at the *positive*, or actually established, only through *negation* or exclusion, at the determination only through the surrender of our free determinability.'[36] And the elaboration continues in the following letter, the Twentieth:

> So it is not enough for something to begin which did not previously exist; something must cease which previously did exist... [One] must in a certain fashion return to that negative condition of sheer

> indeterminacy in which he existed before anything at all made an impression upon his senses. But that condition was completely devoid of content, and it is now a question of reconciling an equal indeterminacy and an equally unlimited determinacy with the greatest possible degree of content, since something positive is to result directly from this condition. The determination which he received by means of sensation must therefore be preserved, because he must not lose hold of reality; but at the same time it must, insofar as it is a limitation, be removed, because an unlimited determinacy is to make its appearance. *His task is therefore to annihilate and at the same time to preserve the determination of his condition*, a thing which can be done in only one way – by opposing that determination with another.[37]

So for Schiller, the sensuous as real can combine with its opposite, the freedom of thought as active, only through the condition of real and active determinacy, and this condition he terms the *aesthetic*.[38] Hegel had much to ponder here, and acknowledges as much in the Introduction to the *Lectures on Aesthetics* decades later.[39] But by then, art is largely divested of its negating power. In the *Phenomenology*, on the other hand, the question of that power was already received as inextricable with creative power, *poesis*, and so Hegel felt it his task to explicate what might be that power's source and identity, as origination.

That explication begins already in the first part of the Preface. If Schiller's 'living shape' combines the sensuous with the formal activity of thinking, Hegel begins his reflections on 'scientific cognition' with a homologous relationship: the coming together of substance and subject. He writes: 'In my view, which can be justified only by the exposition of the system itself, everything turns on grasping and expressing the True, not only as *Substance*, but equally as *Subject*.'[40] But exactly how might these two unite into a whole as True? How does one make the substance *living*, and not simply inert in its immediate substantiality? Or, how does the subject actualise that substance, and in doing so actualise itself out of its own immediacy? Substance can become living, actual being 'only in so far as it is the movement of positing itself, or is the mediation of its self-othering with itself'. It is not enough for one side of the substance–subject equation simply to take hold of the other, and assume it under its own predicating powers. The substance (or the subject) must mediate itself out of its own inert immediacy. But how? How is such self-othering possible? Only through negation, says Hegel. 'This [living] Substance is, as Subject, pure, *simple negativity*, and is for this very reason the bifurcation of the simple; it is the doubling which sets up opposition, and then again the negation of this indifferent diversity and of its antithesis [the immediate simplicity].' This Hegel calls 'the process of its own becoming, the circle that presupposes its end as its goal, having its end also in its beginning'. This process is the 'labour of the negative'.[41]

Negation is first introduced, therefore, as the process of becoming, of coming into actuality. This process requires mediation: negation cannot simply stand inactive in an immediate state of inertia, on one or the other side of the equation, as if in some pure original state. It must penetrate the two sides through movement, and do so as interpenetration. In such interpenetration, the two, as the othering of its respective selves, become in fact identical. 'For mediation', Hegel says moments later, 'is nothing beyond self-moving [*sich bewegende*] selfsameness, or is reflection into self, the moment of the "I" which is for itself pure negativity or, when reduced to its pure abstraction, *simple becoming*.'[42] Such reflection into itself, as self-division, is the great mark of Reason as self-generative, or what Hegel, following Aristotle, calls *purposive activity*. To move out of Substance towards Subject, or from Subject to Substance, one must be self-moving (*selbst bewegend*). This 'power to move, taken abstractly, is *being-for-self* or pure negativity' where 'the result is the same as the beginning'.[43]

This circular movement is possible only by means of a negation always in operation. If this is a *circulus vitiosus*, it is not so because it is self-confining and self-enclosed. For negation, by its very nature, dissolves the seal of the circle, and though it continues in a circular movement (substance ↻ subject), it continually creates new circles. Thus Hegel will later say,

> The circle that remains self-enclosed and, like substance, holds its moments together, is an immediate relationship, one therefore which has nothing astonishing about it. But that an accident as such, detached from what circumscribes it, what is bound and is actual only in its context with others, should attain an existence of its own and a separate freedom – this is the tremendous power of the negative; it is the energy of thought, of the pure "I".[44]

This latter assertion – the tremendous power of the negative – appears in a paragraph that has now, as we'll see further below, become a definitive paragraph of Hegelian negation in contemporary thought, for it goes further and calls us to tarry with this negative in all its power: the living Spirit, the Substance made Subject, is living by virtue not of shrinking before the death inherent in its own dissolution, but of facing it head-on. 'It wins its truth only when, in utter dismemberment, it finds itself. It is this power, not as something positive, which closes its eyes to the negative... on the contrary, Spirit is this power only by looking the negative in the face, and tarrying with it. This tarrying with the negative is the magical power that converts it into being.' But the power is not really the stuff of magic, for Hegel quickly tells us it is the Subject itself, whose immediacy is sublated as 'authentic substance, that being or immediacy whose mediation is not outside of it but which is this mediation itself.'[45]

Introduced here as the self-motivating power through dissolution, the negative is not simply what allows conversion, but, in its most basic operation, allows creation into being. For the conversion of a thing's innermost being to its innermost otherness is in fact a creation of a new being that did not exist before. Such a creation, if it is to operate first within the self-enclosed circle of being, requires a void, or a voiding. But it is precisely this void, as voiding, that initiates the creation. Hegel, still in the Preface, writes: 'That is why some of the ancients conceived the *void* as the principle of motion, for they rightly saw the moving principle as the *negative*, though they did not yet grasp that the negative is the self.'[46] So we can see here that the movement to follow, the movement that will drive the entirety of the *Phenomenology* forward, the movement that, we have claimed, is the power of art, is now, also, and necessarily, at its innermost core, a negation. But a *creative* negation, an art of negation, which now spills forth through the rest of the Preface and indeed the entire argument that rests upon it. The Preface is just the beginning, but if we take it at its word, that beginning is also its end. And the end is also the beginning.

In the development of the argument of the *Phenomenology* proper, we see this movement of negation at work in various capacities and guises. Hegel will use different terms in the opening stages of the text to describe or nuance this force: *difference*, *pure opposite*, *antithesis* (Entgegensetzung), *contradiction*, *self-sundering* (Entzweien), *division*, and indeed *negation* itself. For example, in the description of perception, with the object as Thing in the sensible world: 'The Thing is posited as being *for itself*, or as the absolute negation of all otherness, therefore as purely *self*-related negation; but the negation that is self-related is the suspension of *itself*; in other words, the Thing has its essential being in another Thing' (§126). Or in a subsequent passage on appearance and the supersensible world, concerning force: 'Force, as *actual*, exists simply and solely in its *expression*, which at the same time is nothing else than a supersession [*Sichselbstaufheben*] of itself' (§142).[47] Or again, 'This self-identical essence is therefore related only to itself; "to itself" implies relationship to an "other" and the *relation-to-self* is rather a *self-sundering*; or, in other words, that very self-identicalness is an inner difference' (§162). Or in the subsequent section on self-consciousness: 'Consciousness has for its object one which, of its own self, posits its otherness or difference as a nothingness' (§176). And all this culminates in the final section on Absolute Knowing, where, drawing on the notion of incarnation and kenosis, Hegel speaks of the Self as absolute Spirit and Notion

(*Begriff*), which takes it shape, after surrendering its eternal essence, in the real world: 'The *self-sundering* or stepping-forth into existence stems from the purity of the Notion, for this is absolute abstraction or negativity... It is only through action that Spirit *is* in such a way that it is *really there*, that is, when it raises it existence into *Thought* and thereby into an absolute *antithesis*, and returns out of this antithesis, in and through the antithesis itself' (§796).

From this central movement, whose various segments might expand outward to reveal further vectors of the negative at work, we can move forward to the section on Religion, and now modulate the understanding we have been exploring above of why Hegel couples art and religion together. Both art and religion are expressions not only of the self-begetting that shapes itself through art into religion, but also of that contradictory urge or activity, that self-sundering, that allows self-consciousness to come into existence against its own interior difference. When Hegel says of the Religion of Art that the 'Spirit has advanced from the form of *Substance* to assume that of *Subject*, for it *produces* its shape, thus making explicit in it the act, or the self-consciousness' (§748), act and self-consciousness are here equated, but only through an incarnational unity of substance and subjectivity, a unity that is ultimately, as the Preface laid out in advance, a negativity that abstracts both sides into 'absolute Being' (and along with it both essence and existence, both life and form, etc.). Thus the later language of *kenosis*: the '*self-sundering* or stepping-forth into existence' that stems from an 'absolute abstraction or negativity' (§796).

Now if the power that allows movement forward in creativity and activity is also a negating power, because origination stems from the *Entzweien* (literally, the splitting in two) of what was once a whole or a unity, or from the negation of pure identity into difference, and from the negation of that difference back again into a self-conscious unity, then the inherent dynamic that shapes art and religion, and art *as* religion, is much more primordial than what the outline of the *Phenomenology* might first suggest. If we follow Schiller's cue, as surely Hegel had done, and see art not first as aesthetics, an *a posteriori* reflection on the nature and work of artistic practice, but as an *a priori* impulse or drive that sublates two opposing forces, then we gain a better sense of how art might function in the *Phenomenology* fundamentally: not as a categorical practice in and of itself, but as a shaping force that underlies our most significant activity and formations, individual and collective. The art of 'Art as Religion' would thus become much more than simply stages along life's way, ending in philosophy. Art would be the creative impulse at its most primordial, figuring in the very movement upon which the *Phenomenology* is based, the originating movement behind the back of consciousness. Art could not receive its own category, since it could only emerge as a manifest expression, a *Gestalt*, when that movement becomes 'perfectly transparent to itself', and this, for Hegel, only happens in the moment of Religion. But this means that Religion, and indeed Philosophy, the Absolute Knowing to follow, become creative expressions bound to this impulse, or creative expressions *of* this very impulse. And yet the fact that the impulse itself is a *negative* impulse remains so disruptive to our thinking about art, religion, or even philosophy.

It seemed disruptive enough for Hegel himself, who did not maintain the explicit notion of art as a shaping impulse in his subsequent thinking. In fact, we know that by the time of the *Lectures on Aesthetics* of the 1820s, art had become the starting point of a triadic movement whereby it must, in the end, concede to, even in a self-consuming manner, religion and philosophy. And for many this leads to Hegel's now (in)famous idea that art becomes 'a thing of the past'. But here in the *Phenomenology*, art, in the transformed sense we have now given it by way of Schiller, is a 'death' only as a negation that, at the same time, is a creation, an annihilation that is an origination. It is not that art gives way to religion and philosophy, or even that Art as Religion gives way to Revealed Religion, which in turn gives way to Absolute Knowledge. Rather art is an enduring impulse, one that, as a negating dynamic, originates the movement of the act of consciousness. Art is the emergence of a productive force on the very site that has remained too often obscured in various Hegelianisms since Hegel: the creative and effectual power of the mind as a self-sundering act.

Thinking of negation as the impetus or *conatus* of movement's vitality is not only, as we will further discover, something at the heart of the *poesis* that both brings art into being and then perhaps leaves it behind (in its 'death', or, as Agamben will say, without content); it is also that the very nature of this movement must be understood from the question of the creative activity of art *in the first instance*, before it ever reaches the system (i.e. aesthetics). The *Phenomenology* becomes an important first text in this regard. If art after the *Phenomenology* has, on one end, played itself out as something surpassed (even to the point of death), then the *Phenomenology* allows us to retain or regain a view of art that remains originary, before it has been sundered fully from religion and indeed philosophy. But we should go further and see this originary moment of art *before it is art*, before it takes actual shape in the expressions we call art, or what might amount to the same thing, in Hegel's terms, *after* it is art (after it has been surpassed as a shaped expression). If this originary moment has any effective power, then religion and philosophy themselves become affected, or infected, and 'surpassed' themselves.

What would this 'surpassing' mean, exactly? For Agamben it means surpassing the 'swamp of aesthetics and technics to restore to the poetic status of man on earth its original dimension'.[48] Here aesthetics means art either as reflective technique or as technics, even technology (in the Heideggerean sense). Either way, it is turning the creative impulse into the system that later became Hegel's *Aesthetics*, in which Hegel says the greatest need for the day is not creating art again, but 'knowing philosophically what art is'.[49] To go beyond this kind of aesthetics, which loses the force of art to a purely reflective stance, and makes it into a 'fine art', void of any content or action (and let us remember the sense of finality in the word 'fine' of 'fine art'), is to go back to the *Phenomenology* and retrieve art as an integral feature of the coming to being of Spirit. And if we push this further, this surpassing, which is actually a return – a return that evokes Nietzsche's circle of eternal return, which Agamben picks up in relation to Nietzsche's conception of art as 'the highest task of man, the true metaphysical activity'[50] – this surpassing surpasses even religion and philosophy as similar systems of static reflection. This would mean religion and philosophy would become more than what any one fixed system could maintain, more than the mere institutionalisation of thought and practice, more than compartmentalisations of experience. It would mean religion and philosophy, like art, would always be moving beyond themselves and into the other. The beauty of the *Phenomenology* is that it is a wholly organic expression always on the move. Art is religion, which is philosophy; and even though 'Absolute Knowledge' may be the apotheosis of the Spirit's movement, we can see its own negative organicism will always disrupt that apotheosis in a kenotic emptying. The dialectical movement, in all its so-called progressive moments, is never fully a movement toward absolute synthesis, then. It is a movement driven by absolute negation in which synthesis is only a stage made necessary by negation, and yet always falling to negation, though falling in a productive manner as it yields to the movement that is negation.

The movement of the *Phenomenology of Spirit* as a text thus takes us beyond the traditional categories and limits of art, religion and philosophy. As we have said, it reverses Kant by breaking open limits. It takes us beyond fine art, beyond fine religion, beyond fine philosophy. It keeps them 'unfinalised', moving together toward their own transcendence, yet entwined one within the other. In this movement, we are moved ourselves into a wholly new space in which art, religion and philosophy must be radically reconceived, even beyond the later Hegel himself. Under such an impulse the *Phenomenology* can be called a work of art. Not because it is a fine thing of beauty. Not even because it shows the genius of a creative mind. But because its own productive impulse works against its self to make something new of itself. And in this sense Jean-Luc Nancy was right to call Hegel 'the inaugural thinker of the contemporary world'.[51] For if modernity, or postmodernity, or post-postmodernity, is to surpass itself, as it continually strives to do, in spite of itself, it will do so only by reconceiving itself in its own creative negativity.

CHAPTER 2

Negation's Logic in *Science of Logic*

If negation is a force with a creative nature, we now must look more closely at the origination of that force as itself an originating force. The variant forms of negation we had seen in the *Phenomenology of Spirit* are a precursor to the negation that will be worked out more specifically, and more radically, in the *Science of Logic* (*Wissenschaft der Logik*) of 1812.[1] It is true – we do not normally associate art with logic. But in going to this text, though we may temporarily leave aside the question of 'art' directly, we will nevertheless be looking at the more foundational features that we have claimed constitute art as an originary movement.

In the Preface to the First Edition (1812), Hegel calls the *Science of Logic* a sequel to the *Phenomenology of Spirit*. By this, Hegel was picking up on his claim in the *Phenomenology* that consciousness, if it is to be understood in its most essential form, in its most active formation, and throughout all the modes of its expression, is 'thought which begets itself'. In the 1812 Preface, this phrase becomes 'spirit as a concrete knowing', a spirit that has gone out into the external world, but also a spirit that then frees itself from 'external concretion', towards what in the earlier text was an absolute knowing and now here is a pure knowing, which ends with an absolute idea. The *Phenomenology* ended with the *Begriff*, the Concept or Notion[2] as the pure thought that does not only beget itself as thought but goes further and *knows* itself as self-begotten thought. In order to extrapolate this *Begriff*, the next phase of the journey, one must attend to the pure essentialities of knowing, of 'spirit thinking its own essential nature'.[3] Spirit, as world self-consciousness, must then give over to *Begriff*, and in that shift one must attend to the formations of logic, which are constituted by the pure essentialities of knowing that is *Begriff*.

THE BEING OF LOGIC, OR, THE LOGIC OF BEING

But right away, in the Introduction, Hegel reconceives the notion of logic. The received understanding of logic had been as an abstracted form of cognition, or a set of abstracted forms, to which subject matter is subsequently supplied. But this understanding still implies a separation of the Kantian and pre-Kantian kind: our forms, even if determining of subject matter, are nevertheless dependent upon that matter as extrinsic reality outside them. Hegel wants to show us now that logic, as a feature of absolute or pure knowing, cannot be divided in this manner. Logic is not formally empty, filled subsequently by content from without. Logic is self-filled, with a substantial content of its own. The substantiality of this content brings the inner and outer together in a unity – *form and content*. And here Hegel resorts, once again, to the metaphor of organic nature. When logical forms 'are taken as fixed determinations and consequently in their separation from each other [form from content, phenomena from noumena, finite from infinite, etc.] and not held together in an organic unity, then they are dead forms and the spirit which is their living, concrete unity does not dwell in them.' In order to quicken them, to unify them in an organicism, and fill them with substantial living content, one must see that 'logical reason itself is the substantial or real being which holds together within itself every abstract determination and is their substantial, absolutely concrete unity.'[4]

Logic is 'real being', then. But how is this possible? How does logic escape pure formality, and become living being? Or, how can the structures of thought be seen as synonymous with an ontology? To understand how Hegel sets up his reconception of logic at the beginning of his great tome, let us dwell for a moment on this idea of a unification between form and content, for this will allow us to frame these questions in the context of a mode of art.

The distinction between form and content is often best illustrated by examples from the world of conventional artistic practice. Consider any aesthetic mode of expression within the 'arts': literature, theatre, painting, music, dance, film, etc. In each of these generic modes there are subcategories of different specific modes. So for instance in literature we have numerous different genres, and within each genre numerous different forms: poetry breaks down into epic poetry, lyric poetry and so on, or novels break down into epistolary novels, historical novels and so on. We generally understand the form to be distinct from the content: there is a story to tell, and it is up to the author to find the best genre, form or mode of expression to relate that story. Or in visual art, there is an image to portray, and it is up to the artist to find the best way of visually expressing that image, given the variety of forms available and of materials associated with each form. But art (good art) has always taught us that there is a very close relationship between form and content. Traditionally, a story about the gods needed to be told in a particular way – through poetry, largely. In the Middle Ages, the image of the Bambino had to be painted in a particular manner or style, as opposed to, say, a hunting scene. With the onset of modernity, this relationship became more than merely conventional. If the artist now had an inner self to express, then how he or she would express it became much more elided with what he or she expressed. Charles Taylor has made much of this in his idea of an 'expressivist turn' around the time of Kant and Hegel. In the example of the novel or the play, he says that 'the expression will also involve a formulation of what I have to say. I am taking something, a vision, a sense of things, which was inchoate and only partly formed, and giving it a specific shape. In this kind of case, we have difficulty in distinguishing sharply between medium and "message".' Taylor then concludes: 'For works of art, we readily sense that being in the medium they are is integral to them. Even when it is clear that they are saying something, we sense that we cannot fully render this in another form.'[5]

As modernity advances to the modernism and postmodernism of the last century, we see this relationship pushed to an extreme. It is no longer the case that the realm of divinity deserves a form of grandeur like the epic, or that (to use Taylor's example) Richard Strauss individuates *Also Sprach Zarathustra* through his tone poem, and makes it wholly different to Nietzsche's original. It is that form *becomes* content, and content form, in a consummate unity or indistinguishability. The colours and canvas of one of Rothko's paintings are trying neither to point to something outside themselves, nor even to individuate a certain feeling or sense or idea within Rothko. The form of the painting is nothing other than the content of the painting. What the painting is saying is nothing more than the form it expresses as painting. Hence the genre of 'abstract expressionism' – rightly understood, even the so-called interior expression of the artist is 'abstracted' into the form of the material used, so that there is very little abstraction in the sense of moving from the material towards the non-material, but rather the material and non-material become one and the same thing.[6]

Now it is precisely in this manner that Hegel wants us to understand the working of logic. Rather than merely a formal mode of expressing thought, by means of certain principles, rules and procedures (non-contradiction, syllogism, etc.), or by means of categories (as in Kant's famous *a priori* categories), through which or by which or to which we import content drawn from elsewhere, Hegel wants to show that logic, constituted by the essentialities of *Begriff*, supplies its own content, and that this content is precisely the form itself. 'Hitherto,' says Hegel, 'the Notion [*Begriff*] of logic has rested on the separation, presupposed once and for all in the ordinary consciousness, of the *content* of cognition and its *form*, or of *truth* and *certainty*'.[7] But now, he avers, we must think anew. If consciousness is thought that begets itself, then the structures of that thought must be such that, in the *Begriff*, it *knows* itself as thought that begets itself. *What* it thinks is precisely *that* it thinks, so that the object of its thinking is also the subject as thinking, or the content is the form. So Hegel says, in connecting back to his earlier work: 'Absolute knowing is the *truth* of every mode of consciousness because, as the course of the *Phenomenology* showed, it is only in absolute knowing that the separation of the *object* from the *certainty of itself* is completely eliminated'.[8] The pure science of knowing thought thus eradicates the subject/object dichotomy, so problematic in modernity's pre-Kantian and Kantian legacy. As science (*Wissenschaft*),[9] it 'contains *thought in so far as this is just as much the object in its own self, or the object in its own self in so far as it is equally pure thought*'.[10]

Now, as Hegel suggests above, we should have been able to deduce this unity of form and content from the *Phenomenology* itself, from the modes of consciousness as it moves towards absolute knowledge. So that by the time we address the *Begriff* and its essentialities of knowing in logic, the transfer should be relatively straightforward: thought that begets itself as thought, and is conscious of itself in pure self-consciousness, must now become thought that knows itself fully in that begetting, and knows what that begetting entails in its very essential structures.

But Hegel goes yet further. It is not just that, in logic, the form of our thinking is also the substantial content of that thinking. One could argue that, *mutatis mutandis*, such fusion was inaugurated in the Kantian critical enterprise. Nor is it just that the subjectivity of thinking erases its opposition with the predicate of objectivity. It is also that the unity of form and content, of subjective and objective, is to be in some sense *living*. That is, it must be *organic*, capable of growing and producing in a unificatory manner, and of animating itself beyond empty formalities on the one hand and insubstantial content on the other. In the end, logical thought, Notion, *Begriff*, must be tied to *being*. And this, as we have said, is what Hegel sets out to do, equate logic with living being. That is to say, the modalities of thought in logic become the very same modalities of being in its ontological reality.

This side of Heidegger, to assert that our structure of thinking is isomorphic with our structure of being, or better, that they are one and the same structure, may not seem so extraordinary a claim. But well before Heidegger, and very much influencing Heidegger in this regard, Hegel's claim here is exceptional, at least since medieval and modern philosophy.[11] For it takes the question of thinking ('what is thinking?') and makes it answerable only in terms of the nature of being. It is not deductive as in the Cartesian construction of the *cogito*: thinking leads to the conclusion of being, or fundamentally underwrites it. Nor is it in the conditional as in the Kantian scheme: thought presents to us nothing more than the possibility of things.[12] It is organic: thinking *is* being. As Hegel says, 'the absolute truth of being is the known Notion [*Begriff*] and the Notion as such is the absolute truth of being.'[13] And the 'is' here in its copulative sense equates comprehensively and without remainder.

Yet Hegel goes *still* further, as if this was not enough. The 'is' does not merely equate; as organic, it also, copulatively, *engenders*. This generative aspect of its nature is probably the most difficult to comprehend or accept in Hegel, and is what many Hegelian scholars have failed to appreciate, if they have understood it at all. Yet it is what most sets Hegel apart from others on the point of thinking (logic) and being (ontology). *The copulative 'is' also begets*. This at first seems impermissible. For when we say something *is* something we normally presume to operate under the syntax of identity – what lie either side of the connecting verb are identical, grounded on the truth of equative being (even if that ground is, onto-theologically, God). We do not introduce a generative function, whereby one side gives birth to the other. To say that 'is' also begets is a logical travesty, for how could identity suddenly produce? And what would it produce if not, tautologically, its own self?

If we think of this generative function again in terms of form and content, we can see a development unfold. We first began with logic that, as a form of thought, fills itself with content that is extrinsic to it. Hegel then tells us that we must think differently: a reconceived logic, that is, a *speculative logic*, supplies its own content, so that what is the form is also the content. He then suggests that such content, as intrinsic, must be living content: it must pertain to the very being of the thinking activity. But then the final straw: to be living, it must also engender itself. That is, form does not simply become identical to content, and vice versa; *form brings its own content into being*, and vice versa. In speculative logic there is proper *formation*.[14] Thus Hegel at the end of the Introduction to the *Science of Logic* says, 'The form, when thus thought out into its purity, will have within itself the capacity to *determine* itself, that is, to give itself a content, and that a *necessarily* explicated content.'[15]

This 'content' Hegel will go on to explicate, but in doing so, he is explicating also both form and formation. Or, to put this another way, and in line with the *Phenomenology*, the explicating itself becomes the explicated. And thus it remains 'living'. It is not that logic somehow transcends the function of 'in the now' formation; this Hegel saw as one of the lingering problems of Kant's understanding.[16] It is that the form as formation must be accounted for *in its very formation*, so that the transcendental comes back down to earth, so to speak – that is, to being. In doing so, the objective and subjective unite, both within the individual and between the individual and the world. In the first case, one's consciousness or ego is no longer objectified transcendentally, as a mere instrument for being conscious *of something*,[17] but now looks at the agency (subjectively) as part of its very own something (objectively).[18] In the second case, the subjective ego, as a finite determination, no longer places the object, as infinite other, outside itself. Stephen Houlgate describes this latter point this way: 'The logical structure of the concept "something" – a concept that *we* must employ – is at the same time the logical structure of whatever *is* something in the world. The concept of "something" is inseparably linked to that of "other", "being-in-itself" and "being-for-other" and, correspondingly, whatever *is* something in the world is also inseparably related to what is other than it.'[19]

We might see this unity better in terms of the famous dual nature that accords to the preposition 'of' in its genitive function. (This dual nature, significantly, and as far as I am aware, operates in any language, even those of an inflected nature. As we will see, this universality might have something to do with the genitive case itself, as a case that not merely possesses but, at its root, begets.[20] So too, in the very semantic thrust of the preposition 'of' we find the sense of beginning, sourcing, proceeding, especially in relation to movement.) For example, in the phrase 'consciousness of something', we normally understand the prepositional force as an objective genitive: consciousness holds in its grasp an object that is something – possessively, 'consciousness's something'. We would only understand the preposition subjectively in the context of working out that something's essential nature: if we think that something might have consciousness, we would refer to the consciousness of something with a subjective genitive – possessively, 'something's consciousness'. According to Hegel, in the very relation to consciousness or ego, Kant was only interested in the objective construal. (And this is what kept the ego 'transcendental'.) But Hegel does not correct Kant by merely going to the opposite, the subjective construal. Such a move would only retain the fundamental opposition in play. Rather, he means to include both construals, both forces – *at once*. So 'consciousness of something' is both 'consciousness's something' *and* 'something's consciousness'. Both subjective and objective. This bi-directionality, as deployed by Hegel, we might term *cogenitive*.[21] The genitive force is mutual, co-extensive. In the cogenitive the preposition ('of') becomes proposition ('is').[22]

To explicate this cogenitive function, Hegel will show that the 'logic of being' operates in precisely the same way: it is both logic's being *and* being's logic. For the relation between 'consciousness of something' and 'logic of being' (and hence, we might say, between the *Phenomenology of Spirit* and *Science of Logic*) requires that, in order to treat consciousness *qua* consciousness, and to allow it the power to transcend its own transcendence, one must introduce the *Begriff* as thought thinking about thought *as such*, and not as thought *of something else* (objective genitive). In Hegel's words, still countering Kant, 'But if philosophy was to make any real progress, it was necessary that the interest of thought should be drawn to a consideration of the formal side, to a consideration of the ego, of consciousness *as such*, i.e. of the abstract relation of a subjective knowing to an object, so that in this way the cognition of the *infinite form*, that is, of the Notion [*Begriff*], would be introduced. But in order that this cognition may be reached, that form has still to be relieved of the finite determinations in which it is ego, or consciousness.' From which then follows the sentence we have already quoted: 'The form, when thus thought out into its purity, will have within itself the capacity to *determine* itself, that is, to give itself content.'[23] Thus we might now add the cogenitive phrase: 'the form of content'.

Now all of this, difficult as it may be, is by way of a prelude – we have not yet left Hegel's Introduction to the *Logic*. Why have we dwelt upon it? Because, for one, it is crucial to see the living nature of logic as being. But it is even more crucial to see the generative nature of this logic. And both are captured in the cogenitive nature of the phrase 'logic of being'. For in order for logic's being to become, in a circular move, being's logic, identity must be *generated*. The pre-positioning of 'of' (X of Y) must turn into the copulative identification of 'is' (X's Y is Y's X), which in turn must engender a pure identity (X = Y). 'Is' must ultimately give birth. But how? How does 'is' give birth to something without destroying itself, at least in the copulative sense of mere equation? How does being become cogenitive, the proposition synonymous with the preposition? We are now at that position whereby we can introduce the first concept that will mark the 'explication' of being in the first main section of the *Logic*, negation.

But how does one begin?

THE PRINCIPLE OF BEGINNING

From the foregoing, we could claim that Hegel is still on a journey, the journey begun in the *Phenomenology*. On this journey, consciousness moves to self-consciousness, whose grand destination is Absolute Knowing, which is constituted by *Begriff* (Concept or Notion). *Begriff* then requires a *Wissenschaft*, even begets a *Wissenschaft*,[24] a knowledge that can appropriate its very knowing, and explicate the essential structures of that knowing. This knowing, as formal thought, will necessarily entail being, since knowing must be constituted, even before essence, by the act, as living activity. Being, then, must be the first point of departure for this next stage. But being, as objectively understood *qua* being, will lead to essence, and essence, as the truth of being, will lead back to *Begriff*. Moreover, the return to *Begriff* will be subjective, since it is a *Begriff* that has 'withdrawn into itself from externality', and thus it will lead back to the Absolute Knowing in which consciousness, as self-consciousness, culminates. So we might see the journey as circular, even in the cogenitive sense (the *Begriff* of *Geist*). Indeed, the final line of the entire *Science of Logic* speaks of *Begriff*'s completion within a '*science of spirit*' (*Wissenschaft des Geistes*), which returns us back to the *Geist* of the *Phenomenology*.[25]

But as we will see in the explication of being, the concepts of the linear and the circular ultimately are not, as traditional logic would tell us, mutually exclusive. For Hegel, what returns, what loops back in a circle, is also what propels in a forward movement. And vice versa. It is the same problematic as 'returning anew' – the co-residency of opposing conditions.[26] Here, in the opening section on being, we find Hegel's greatest elaboration, and justification, of this paradox.[27]

Being – what is it? Hegel wants to be sure that we think being first without any prior conceptions or prejudices. Here he looks back to Descartes (one must create a *tabula rasa* about the understandings of certainty we have inherited), and forward to Heidegger (one must try to recover the sense of being buried for so long by the Socratic/Platonic tradition). And so Book One of *The Science of Logic* begins with questions of beginning. Where does philosophy, or the philosophy of being, begin? Or how? Do we ask, Who has handed down to us the proper concepts with which to begin? Or do we ask, What is the principle of beginning in philosophy, before we attach any names to it? The first involves that which is 'mediated', the second 'immediate'. To rephrase the matter, we could ask: Is the question 'what is the beginning of philosophy?' an objective one ('what is philosophy's beginning?') or a subjective one ('what is beginning's philosophy?')? Or: Is the *principle of beginning* a question about principles as such, or beginning as such? Hegel tells us that, traditionally, we have favoured the objective side: the principle, in the question of beginnings, has been construed by means of determinate content. What stands at the beginning (of everything, and thus of philosophy) is some content that is either concrete (water, substance, etc.) or conceptual (the one, *Nous*, idea, monad, etc.). Or if it is the nature or process (form) by which content comes to us ('thought, intuition, sensation, ego, subjectivity'), still it is the content which has remained 'the point of interest.'[28] No one, says Hegel, asks the question, 'With what should the *beginning* be made?'[29] What is the ground of origination? But to ask this question, which addresses the subjective side of beginning (not the beginning of something else, but the beginning of beginning), requires that we put logic to the test. For normally conceived, the point of departure about beginning evades logic altogether: one starts with a revelation, or with faith, or with some form of intuition (even if intellectual). But all of these departure points leave aside the question of method inherent to logic. The *Wissenschaft* of logic then must begin with the question of beginning for its own sake, and show that it can withstand the methodology of logic. But to do this, we must think differently about logic as much as we must think differently about being. Hegel's organicism emerges yet again. Hegel says, 'Thus the principle ought also to be the beginning, and what is the first for thought ought also to be the first in the *process* of thinking.'[30]

This in itself is a crucial departure point, for the principle (the principle of beginning, cogenitively) is not just a principle that we begin with ('let's start with this principle, for it is the furthest back that we can go'), but the beginning as such (let's start with the idea of beginning as the first principle). This places the question of beginning *as origination* at the core of the matter.[31] How does something come *to be*? Here our thematic of art will gain its potency towards the reformulation we have been suggesting. This is a potency that goes all the way back, as Heidegger rightly saw, to the Presocratics, and here it is worth looking again for a moment at the ancient term *arché*.

The *arché*, translated most often as 'beginning', was, for the ancients, more than just a point of departure. It was a primary or principle cause or origin, or the principle of origin itself. Although Aristotle in the *Metaphysics* defined it as the 'first point whence a thing's movement proceeds', he quickly acknowledges that common to all beginnings is 'that they are points of departure either for being, or becoming, or knowing.'[32] And so in referring to the earliest of the Milesian Presocratics, Thale's famous *arché*, that of water, becomes the material principle or 'primordial being' that acts as a founding substrate of all things.[33] Other Milesians had other *archés*: Anaximenes' was air, while Anixamander's was the *apeiron*, an indefinite primary substance. Now whether these Milesians themselves utilised the term *arché*, or utilised it in the way that Aristotle understood it, has been debated.[34] But that Aristotle at least saw this principle of *arché*, however one fills its content, as inherently related to being, to becoming and to knowing shows that *arché* itself, as a general principle, is one that undergirds both ontology and epistemology. Like the cosmogonies of his predecessors, going all the way back to the mythologies of Homer and Hesiod, the *arché*, as beginning, concerns the bringing into existence, or the bringing to be as being (*on*), but now in a philosophical sense it must also account for the origin of origination itself. Aristotle will work this out in the kinetic terms of movement ('that whereby the movement begins');[35] but his predecessors had already seen that the notion of principle itself is synonymous, cogenitively, with the notion of beginning. And however this unity of synonymy is manifested – even if, as Aristotle noted, it is manifested in contraries, as it was for the Pythagoreans[36] – it is a unity that is properly the beginning of philosophy.[37]

Hegel well understood this too. Just as Aristotle sought the need for a proper mode of explanation for any beginning as *arché*, a logic for the principle, as it were, so too the beginning of any *Wissenschaft* of logic must begin with logically grounding 'beginning', or with beginning grounding logic, a circularity that acts with the same cogenitive force as above, the principle of beginning, which is to say, the principle *as* beginning (*arché* in its immediacy).

But Hegel also knew perfectly well that one cannot grasp the principle of beginning in pure immediacy alone. What would this look like, pure beginning? How could we isolate it, mark it out? Beginning, like any principle, is also always mediated. Hegel's own text is a great example: the beginning of *Book One* is where it arises. Now it is true that, in a Cartesian sense, Hegel wants to rid us of our presuppositions concerning the beginning. He wants us, in anticipation of the Husserlian phenomenology, 'simply to take up, *what is there before us*'.[38] But we can do this only through the distinction that what is there before us has been separated off from what is mediated (say, in Hegel's case, the *Phenomenology*). Thus simple immediacy, as a beginning – not only beginning with simple immediacy, but also simple immediacy as beginning – is only understood or represented 'as having come to be through mediation'.[39]

But now this mediation must be sublated, or, more properly, mediation 'is also a sublating of itself'.[40] For to return to immediacy, by way of mediation, requires that mediation destroy its own pathway. So we can begin with beginning, which is immediacy itself, but only once we have travelled the journey through the mediation of consciousness to knowing.[41] Once there, however, we must then return to a presuppositionless immediacy, as if we had never travelled in the first place, and this, Hegel tells us, brings us to being. This paradoxical journey is what Hegel calls 'logical beginning [*logische Anfang*]'.[42] And such a beginning 'is pure being'.[43]

THE BEGINNING OF BEING; THE BEGINNING OF NOTHING; OR THE BEING OF NOTHING, AS BECOMING

Let us now, finally, look at what for Hegel is entailed in pure being. If we take everything we have said so far, and apply it directly to being alone, we must first look at being without prejudice, in its immediacy, and for its own sake. But what do we find when, through mediation, the mediation of the cogenitive ('the immediacy of being'), we arrive at immediate being? The problem is, nothing. For here, being as such is also immediacy as such. And what do we have with pure immediacy? It is the same as pure beginning – nothing. Nothing, that is, with which to determine anything. And therefore nothing *itself*.

Hegel's description of the beginning of being is in terms of determination or determinateness (*Bestimmung* or *Bestimmtheit*). The German words here stem from the notion of rightness or correctness (*stimmen* – to be right, true): that which is determined has been set right or certain, made definite or particular through ascertainment.[44] The English translations have generally used the Latinate term, which refers, spatially, to the terminus: that which is determined has a distinct endpoint or boundary, which gives it definition or distinction.

But pure being, as beginning, is *in*determinate, says Hegel. It has no boundary to give it definition or distinction. In spatial terms, it is *in*finite. In mediatory terms, it is *im*mediate, without any *other* through which or by which to set itself apart. It is all-encompassing. Or we might say (to adopt theological language) it is pure plenitude, except that it is filling nothing apart from itself. It is therefore also pure emptiness. In one sense, then, it is *everything*. But if that everything is only itself (it is not a manifold, because it carries no distinctions within itself), then in another sense, it is also nothing. For what is something that cannot be determined, because it carries no determinations? It is, by logical extension, nothing.

Hegel needs only one paragraph each to describe pure being and pure nothing in these terms. Little elaboration is required: if we get back to the beginning, in its immediacy, of a comprehensively pure being, we arrive, at the exact same time, at a comprehensively pure nothing. This, implies Hegel, is pure logic. It need not be expounded. But it is also counterintuitive to all logic traditionally constructed, from Parmenides onwards, since nothing is contradictory to everything that is, and therefore it cannot be countenanced.

How then can Hegel, under the ostensible rubric of logic, assert that being and nothing, in all their purity, are actually one and the same thing? Does this not disrupt the very foundations (onto-theological and otherwise) of our very notions of existence?

But we cannot leave things here, in the stability, which in this case amounts to a profound instability, of a pure equivalence: being = nothing. We should now be familiar enough with how Hegel's notion of identity works: it is not merely a *state* of equivalence, but a *making* of equivalence. And this very *making*, this origination, undermines the equating, so that it can never reduce to a stasis of being, but only to a coming to be.

Let us try to probe this matter to its innermost core. We began with pure immediacy, which is all-pervading being. Pushed to its greatest extent, which is to say, returned to its most primordial point of beginning, there would be nothing else. Nothing but being. But because there is nothing else, there is therefore, by virtue of the logic, also *nothing*. But 'also' suggests something else, even if that something is nothing. How can we have it both ways? Either there is only one thing, being, or there is not. So said Parmenides. But the one, pushed to its greatest extent, becomes nothing, which in turn produces two things: one thing and its opposite. Now to say simply that there is still only one thing, because the two are perfectly equivalent, does not overcome the inherent duality. 'Equivalence', as much as 'identity', assumes two things. From whence does this *other*, as second, arise? It arises from the very genesis of its purity. The moment we reach the state of pure indeterminateness, we generate something else, its pure opposite.

Hegel now has to account for this generation. The very moment pure being is reached, pure nothing comes into existence as well. So at this primordial point of beginning, both being comes to be, and nothing comes to be. Or conversely, both being comes not to be, and nothing comes not to be. However we look at it, something *becomes*. Or equally, nothing *becomes*. And all this takes place immediately (now as much in the temporal sense): as Hegel describes it, the one vanishes immediately into the other. Or, 'the hour of their birth is the hour of their death.'[45] This vanishing is also a becoming: 'a movement in which both are distinguished, but by a difference which has equally immediately resolved itself'.[46]

Becoming, then, originates in the unity of being and nothing. But in a following remark to this becoming, as the basic third 'proposition', Hegel quickly points out the defectiveness of this so-called unity, as identity. If the two distinguishable sides, being and nothing, are in fact in selfsame identity, the proposition ('being is nothing') itself implodes, because the proposition, as *proposed*, contains the two things and their distinguishability. So Hegel says, 'we find that it [the proposition] has a movement which involves the spontaneous vanishing of the proposition itself. But in thus vanishing, there takes place in it that which is to constitute its own peculiar content, namely, *becoming*.'[47]

The creation of its own peculiar content carries us back to the thought that begets itself, and to the form (logic) which creates its own content (being). But now the 'content' is precisely 'becoming'. Thus the proposition ('is') propagates, and what it propagates is the very propagation itself ('becoming'). But it does this only by way of its self-imploding, its self-vanishing, or its self-negation. 'Is', by means of its 'is not', gives birth to 'becomes'. Negation, then, *originates*.

It is difficult to isolate this point of origination precisely, and Hegel is up against much in trying to make it clear. Each of his four 'Remarks' that follow the three simple one-paragraph propositions (Being, Nothing and Becoming) in some way tries to come to terms either with the opposition to such thinking, or with the defective manner in which it could be understood, or with its abstract nature, or with its utter incomprehensibility. In the latter category he places the question of thinking the beginning. The problem, as has already been implied, is that for anything to begin, it cannot already *be*. For if it already is, says Hegel, it cannot be at the point of now beginning, even if *just* beginning. The same goes for the opposite: if it is not yet, then it cannot be at the point of beginning, even if *just* beginning. The conundrum follows a long philosophical tradition, going back to Zeno, where the question of ever-decreasing increments (of time or space) continues on indefinitely, or until the conceptual category itself (say, of arrival) breaks down completely. Perhaps in more contemporary terms, we might think of the opposite end of the spectrum to beginning: Blanchot speaks of the instant of one's death, which any one person can never experience. For the moment one is dead, one can no longer experience. One can experience the dying, but not the instant when experience expires from one.[48] So too with beginning: the precise point of that origination seems impossible to isolate without being something else.

It is only by becoming 'becoming' that it is possible, claims Hegel. Being and nothing become moments, and they are moments that are in continual sublation, being cancelled by the other, its own internal opposite, while simultaneously the very reverse is happening to the opposite.[49] 'Coming-to-be' and 'ceasing-to-be', Hegel says. 'Both are the same, *becoming*, and although they differ so in direction they interpenetrate and paralyse each other.'[50] Only becoming keeps them going, by keeping itself going. (Movement remains essential.) And only this becoming allows for, or, more to the point, generates, the beginning proper.

But there is one more consideration to be had. If becoming was the final 'endpoint', the stabilising proposition ('there is becoming') to the preceding contradiction ('being is nothing'), the ultimate *synthesis* of thesis and antithesis, which now stands proudly on its own as the triumph of a long and hard battle, then we would find ourselves in the strange position of proposing a state that can never be a state. Becoming is always, *in perpetuam*, beginning. It cannot propositionally be, as such. If it is to be the becoming *of* something, even if that something is, propositionally, itself, then the cogenitive must come back into play, and the proposition will re-invoke the preposition, putting a subject back into relation with an object (instead of retaining their unity or pure identity). But it can only do this by, once again, destroying itself, where something – this time itself, becoming! – has become other, the distinction into being or nothing. In Hegel's own words: 'Being and nothing are in this unity only as vanishing moments; yet becoming as such *is* only through their distinguishedness. Their vanishing, therefore, is the vanishing of becoming or the vanishing of vanishing itself. Becoming is an unstable unrest which settles into a stable result.'[51]

Now the 'stable result' here is not becoming as such, but determinate being, which can now occupy the subsequent chapter, as Hegel's explication develops through to the further qualities of being in its determinateness. But we can see the crucial move has now been made (a move that, in itself, entails perpetual movement): the stable result – *any stable result* – is constituted by the unstable unrest that is becoming, and this instability, this restlessness (we will explore this term later, as elaborated by Nancy), is the negative as the movement that is there in the beginning as the beginning.[52] 'In the beginning...' – this is the genesis of negation, cogenitively. Any stability, any result, is not merely haunted by it, but motivated by it, even, in this now exceptional sense developed by Hegel, generated by it.

NEGATION PROPER

We have now arrived at the point where negation can come into its own. Let us first remind ourselves of the subtle but significant distinction we earlier made between negation and nothing. Nothing, as pure void, we have just seen in the immediate indeterminate beginning that is pure being. It is not merely *in* it, but, more troublingly, *as* it. But this nothing is not yet negation, the *making* of nothing. The making, or origination, of nothing is what Hegel has tried to disclose for us (out of the necessity of the Aristotelian demand – *from whence?*) in the notion of becoming. For it is not the *fact* that nothing is there at the beginning. It is rather the *beginning* that is the 'fact' of nothing, there as beginning. But this beginning is comprehensible (for where is 'there'?, we might ask) only in terms of becoming. And this becoming can only remain itself when it resists the propositionality of a stable state. To remain itself, then, is to constantly sublate itself. It is this very constancy of inconstancy, this cognitive paradox, that acts as the driver, the motive force, towards any kind of determination, 'the determinateness from which a thing originates in itself'.[53] And it is precisely this motive force we now call, under Hegel's direction, *negation*.

We cannot remain in indeterminacy, nor immediacy. Where are we if we should linger there? If we cried, who would hear us? As within Rilke's angelic orders, we would only vanish. So determination is what everything, every movement, leads towards, the 'stable result' of becoming through the unity of being and nothing. In the second chapter of Book One, Hegel calls general determinateness, in the form of being (*Dasein*), 'quality'. But this quality still retains its two sides, being and nothing. When we isolate the quality of this (determined) being, we have *reality*; when we isolate the quality of this (determined) nothing or non-being, we have *negation*. So the determination of nothing, as non-being, becomes negation, keeping in mind the noun of action that nega*tion* becomes from nega*ting*.

What is extraordinary here is Hegel's own 'determination' to keep nothing constitutive, even in determinateness. We might have said: fair enough, at the level of indeterminateness and immediacy, we can permit the notion of nothing. But since no one can dwell at this level, nothing becomes a moot point. But Hegel won't have this. Determinate being (*Dasein*) retains its two opposing sides. Why? Because of its origination. It cannot become determined except through the dialectical process of becoming, and this is a process, we have just seen, driven by the cogenitive paradoxes (being of nothing, constancy of inconstancy, etc.) that eventually form, or are formed by, negation. So when Hegel talks about negation as a quality of the *Dasein* (as Heidegger later will, in his own manner),[54] he means to keep in force this origination, because it is the very genesis of the determined state. Thus we have the statement: 'Determinateness is negation posited as affirmative'.[55] The very act of negating, when *posited* as an act, brings determination. So we can still have determined nothing, as non-being, as well as determined being. It is only when that act turns on itself – the negation of negation – that we get something, *a* specific determinate being.[56] But the origination of the entire process must not, cannot, be lost: the becoming of the beginning, which is the begetting of nothing.[57]

Now, an almost entire 'explication' of logic still lies before us, as this doctrine of being works its way towards the doctrine of essence, and both doctrines give way to the doctrine of the *Begriff* in the book's second volume. But the core of the matter, in terms of negative origination, is now established. Everything that follows – and a thoroughgoing analysis would take us well beyond the limitations of our present scope[58] – returns back, in some variation or amplification, to this core, this beginning.

So for example, in the middle section on essence (Book Two) Hegel develops the notion of reflection that he had alluded to only fleetingly in Book One.[59] Here reflection, we might say, plays the role consciousness plays in the *Phenomenology*, or anticipates it after the fact. For essence is determinate being that no longer sees its determinateness in immediacy, but rather in mediation, or in self-relation (what Hegel will call 'being-with-itself' (*Sein mit sich*), the last words with which Book One ends).[60] But in order to 'see' this, it must 'inwardise' itself (*sich erinnert*) through reflection, which carries now an element of knowing (as did, eventually, self-consciousness) that is inward knowing.[61] Thus reflection is equated with essence, but with an essence that has not yet unified its inward nature with its outward manifestation (an essence Hegel will call 'illusory'). Reflection too, then, has its own negation, equally well worked out, but fundamentally tied to the negation of (the) beginning, as we might see in statements like the following: 'The movement of reflection... is the other as the *negation in itself*, which has a being only as self-related negation. Or, since the self-relation is precisely this negating of negation, the *negation as negation* is present in such wise that it has its being in its negatedness, as illusory being [*Schein*].'[62] Or, 'becoming is essence, its reflective movement, is the *movement of nothing to nothing, and so back to itself*... Being only *is* as the movement of nothing to nothing, and as such it is essence.'[63]

Or this negating movement is worked out a little later in the concept of difference, in which absolute difference is both difference and identity, a unity which, 'as the essential nature of reflection', becomes 'the *specific, original ground of all activity and self-movement*', which can be further described as 'negative relation-to-self'.[64] And this finds similar expression in the still later concept of contradiction, which perhaps is the most direct expression Hegel gives concerning the fundamental motive nature of negation in its primordial sense: 'contradiction is the root of all movement and vitality; it is only in so far as something has contradiction within it that it moves, has an urge and activity'.[65]

Each of these later permutations in fact deserves its own analysis, to draw out the nuances of its reformulations.[66] But any differences we will ultimately discover lead back to the beginning, the core we have worked out in detail above, precisely because, organically, they are not, in the end, differences at all, but in identity with the negating force that is at their origin (as origination). They will ultimately constitute, therefore, a 'returning anew'.

The Science of Logic remains one of the greatest books in the Western philosophical canon precisely because of its uncompromising attempt to source the most disruptive of all conditions to that tradition: negation that leads to the beginning of nothing. And for this reason, it is one of the most difficult of texts in the tradition. But that difficulty should not deter us from its significance, nor keep us from its engagement, even if the so-called '*Lesser Logic*', *The Encyclopaedia Logic*, makes its own retreat. Nor should we minimise the influence this attempt has had on Hegel's entire corpus, even if any such retreat allows the Hegelian system of the *Encyclopaedia* as a whole and the absolutising of this system in his later works (and lectures) to come to the fore.

What we need to hold to especially from all the *Greater Logic*'s complexities is its originating 'ground'. Negation, we must remember, after all is said and done, and indeed *before* anything is said and done, is a force that begets, in fact the begetting of the very capacity to say and to do. When this force comes into its own, it becomes an art. Not yet, of course, a disciplined practice of crafting something. Nor yet a disciplined reflection on the nature of that crafting and its results. But the origination of that moment when something, as nothing, comes into its own. This art may often remain 'hidden', as Kant had said of the 'art' that was the synthesising power of imagination in the first *Critique*.[67] So to bring it to light, and to see how its origination permeates all of our thinking and activity, we must catapult ourselves ahead to Hegel's *Lectures on Aesthetics*, to show that even as a work of later systematisation, it too succumbs to the power of this originating force as a diremptive, de-systematising agitation moving within even the most beautiful manifestations of our cultural products and creative endeavours.

CHAPTER 3

Art's Negation in Aesthetics

We have been developing a notion of art that is hardly conventional: art as the originary force that allows self-generation and, in relation to the very act of thinking, 'thought which begets itself'. For many, this might remain a misappropriation of the term 'art'. For the *ars* of the ancients first denoted a skill in a particular domain, a practical craftsmanship, whether in a profession or in creative making. There seems little of a practical nature in our discussions so far about the activity of negation. But we must remember that the term *ars* had a wide application, which, if it began as a practical skill, moved quickly into areas of learning and knowledge, of conduct and character (and hence of virtues), and even of theoretical principles. Still today we hear the phrase that philosophy, and even more precisely logic, is the 'art of thinking'. And Hegel himself applied the term to several different contexts: as we have seen, the art of religion in the *Phenomenology*, but also, in his discussion of the Greek world in his lectures as *The Philosophy of History*, the art of politics.[1] Of course, *ars* was also applied to conjuration and trickery, and perhaps on this footing suspicion is erected: the art of negation appears to carry elements of cunning and deceit. But then, we know from Hegel that reason has its own cunning – the ability to interpolate its own purpose or end into objects and their activity without necessarily being present itself.[2] If we speak of an *ars negationis*, it is not only to encompass the entire semantic range of art's application, from the practical to the theoretical (for this range is exactly what Hegel's philosophy wants organically to unify), but perhaps also to speak of a *cunning of negation* most definitively (and cogenitively): the ability of negation to act out its generative 'purpose' in a manner that negates its own presence, and thus cunning's cunning.[3]

Hegel himself turned to the practicalities of art after, we might say, his system had fully developed. The *Lectures on Aesthetics*, or the *Lectures on Fine Art* as they are sometimes called, were compiled from (at least) three different time periods – 1823, 1826 and 1828/29 – when Hegel was at the University of Berlin. He himself did not do the compilation; one of his students took upon himself the task of creating an edition based on Hegel's own notes and student transcriptions.[4] We say 'after his system had fully developed' because the *Science of Logic* had given way to a three-part elaboration that became the *Encyclopaedia of Philosophical Sciences* (first edition 1817), and the *Elements of the Philosophy of Right* was already penned (1820). Hegel himself published nothing more of book length during his lifetime. (Those that have since become books – lectures on religion, on history, on political science, on art, e.g. – were put together posthumously by redactors.) 'System', as represented in these later texts, denotes the inner logical coherence with which the dialectical movement of the *Geist* moves to its consummate manifestation in the world and its history, a coherence that tends to privilege the 'higher term' brought about through the *Aufhebung*. As we saw in the Introduction, such a system tends to be read in doctrinaire terms, whereby the absolute, as Absolute, has fulfilled its unifying role, bringing identity to opposing conditions in such a manner as to usher in an 'ideal world' – not a utopian world of perfected conditions, but an empirical world (nature) populated by organised human activity (the state) that now wholly manifests the aspirations of its own ideas as Spirit (ideal world), as the *Philosophy of Right* concludes.[5]

ART VS. AESTHETICS

The question of art as traditionally understood, art as the creation of aesthetic objects, had for all intents and purposes not figured in Hegel's writing prior to the later lectures. Various works of art crop up, to be sure – Greek tragedy in the *Phenomenology*, most famously – but not in the service of art *qua* art. The pre-*Phenomenology* writings occasionally reference art, but largely in relation to Hegel's philosophical predecessors or contemporaries, and thus in relation to larger philosophical concerns.[6] Religion as art, we saw in Chapter 1, was not concerned about art in any conventional aesthetic sense; art was the 'shaping' or 'forming' of consciousness into religious expression, which with the Greeks especially may have produced a body of beautiful artwork, but not for the purposes of art in itself or of beauty in itself. Art first makes its appearance as its own category (in the non-Kantian sense) in the *Encyclopaedia of Mind*, where in eight short paragraphs (§556–§563) Hegel reformulates the beginning of *Geist*'s movement through its absolute phases, as first set out in the latter stages of the *Phenomenology*: art, then religion, then philosophy. At one level art here is now treated 'aesthetically' – the 'finitude of Art' is described in terms of 'the subject which produces that work, and the subject which contemplates and worships it'[7] – but at another level, a more significant level, any such 'aesthetics' remains in the purview still of religion, and ultimately of absolute *Geist*. For art, in giving external form to the Spirit, eventually must free Spirit up towards its universality beyond the particularities of sensuous material. And 'Beautiful Art', most advanced in its aesthetic nature, allows this more than any other form, precisely because, as we shall see further below, beauty operates through the 'consciousness of free spirit', and only such consciousness, and such freedom, can bring Spirit out of its external prison and into 'the universality identical with the infinite form'.[8] And that infinite form Hegel will call 'revelation', which carries us necessarily to the next phase, revealed religion. So here too, though art has made some advance in its capacity to be 'aesthetic', it still must give way to the higher phase. 'Beautiful Art, like the religion particular to it, has its future in true religion.'[9]

By the time of the lectures on fine art of the 1820s, one might be tempted to think that 'aesthetics', as the experience of artwork and the reflection upon artwork, has now gained its freedom. It can deserve its own treatment intrinsically, without having to yield to religion and/or philosophy. It can finally deal with real works of art – drama, painting, sculpture, music, literature and so on – as produced by real artists – Sophocles, Phidias, Fra Angelico, Mozart, Goethe, et al. But such freedom would be false. And not simply because the 'end of art' is announced in the Introduction (we have seen, and will see further below, that this 'end' is not an 'end'), but because 'aesthetics' as such is not, nor ever has been, Hegel's interest, even after two volumes of lecture material.

Martin Gammon informs us that 'Hegel himself rarely used the term "*ästhetischen*" in his early writings, except in a negative or pejorative sense.' Gammon goes on to explain: 'Specifically, by "aesthetic" Hegel meant a "spectatorial" relation to the image of beauty, which segregates subject and object in an intellectual reflection.'[10] We might say this of all Hegel's subsequent writings as well. For he rejected such segregation at all points, and rejected any such notions of reflection that would result in it. Art was not about, in any way, spectatorship. If beauty was simply about beholding, or being beheld – *aesthesis* – Hegel had no interest. Art, even from the earliest moments of Hegel's thinking, was about a dynamic interplay between the Spirit and its concrete world, an interplay that was generative and not spectative.[11] This understanding he had picked up from Schiller, who had employed the term *aesthetic* in a manner that brought together the various powers of human experience into a unity: the physical, the logical and the moral combined in a totality.[12] When these powers are kept separate, when our relation to any object results in a reflective objectifying, which beholds a work but does not solicit all these human powers as a unified totality, it is not, either for Schiller or for Hegel, a properly 'aesthetic' encounter. It is 'aesthetic' only in the sensuous sense, which for Hegel meant a disunified sense. Thus, at the very beginning of the *Lectures*, when Hegel addresses the term 'aesthetics' in its most basic meaning, as the science of sensation or feelings, he remains 'indifferent' to that science as it has been passed down. Therefore 'the proper expression for our science', he claims, 'is *Philosophy of Art* and, more definitely, *Philosophy of Fine Art*.'[13]

In maintaining the term 'art', then, we hold to something far more integral than the spectatorial/reflective sense of 'aesthetics', and more integral for the *Lectures* themselves, even in their more 'systematised' development. Art is not about the work in its isolation, either as created, performed or beheld. Art is not even about, ultimately, collective works in their division as multiple forms, even though Part III of the *Lectures* is devoted entirely to 'Individual Arts' (architecture, sculpture, painting, etc.). For indeed that section is entitled '*The System of the Individual Arts*', which indicates that a system is already realigning them in their individual division. The operations of such a system clearly show that Hegel is never truly interested in art for the sake of art alone – and not merely the *l'art pour l'art* of later modernism, but also the nature of art as a complete autonomous category, or a theory of art as self-contained aesthetic experience. As Robert Pippin has argued strenuously, 'Hegel denies the autonomy of the aesthetic, or at least its complete autonomy, and this denial is the basis of the claim that art must be considered a social institution linked to the development of the norms and values of a society as a whole, and that it is best understood in terms of its similarities with religion and philosophy and not as autonomous.'[14] Hegel maintains an organic understanding: if art is involved in a process of absolute *Geist*'s unfolding, that process can never fully leave behind its preceding phases, its beginnings. Indeed, as we have been arguing, art enacts a certain kind of beginning, even at the conceptual level, and for conceptual thought. It is not representative of an isolated aesthetic experience, and nor is it representative of isolated reflective thought. It is not representational at all.

This is why Hegel, still in the *Encyclopaedia Logic*, disavows the principle of 'the imitation of nature' in art. To re-present something aesthetically is to return to the segregation of subject and object. The form that any art takes has to be self-generated, not taken from a wholly extrinsic source. More, it has to render any externality its own, what he calls 'the "characteristic" meaningful natureform which is significant of spirit'.[15] In the *Lectures*, he says it more plainly: the aim of art is 'to have its end and aim in itself'.[16] This seeming tautology does not return art to a self-enclosed autonomy, in which an 'aesthetics' then might prosper. Rather, it denotes a self-generative feature inflected throughout all of the 'system' as such. Art does not mimic the world around us as if a reflecting mirror. Art generates, art *shapes* the forms by which generation, and *poesis* itself, is possible. But of course to do this involves something more than just a produced work. As William Desmond says:

> Hegel's divergence from imitation consists in the fact that according to the imitation theory the original comes *before* the image, is preconstituted with respect to it, while for him the original in its true, full form only comes *after* the image, being the reflective result that is brought to fullest articulation in the further development of philosophy. In both cases the real original in its proper form is *beyond* the image – even if for Hegel art can contribute to the constitution of the original in final philosophical form.[17]

But where is this beyond, precisely? It is not a question of after or before; it is a question, we can now say, of negation. For as we have repeated often enough, the end of negation is beginning, and the beginning of negation is end. And this circularity is what art now brings to the 'system', even if the system ends in lectures on art.

NEGATION'S END

Now the presence of negation does not seem, on the surface, to operate in any constitutive way in the *Lectures*. In the Introduction Hegel begins by asserting what we might have already and easily adduced from the system as it has developed: that art is a middle term between pure thought and sensuous external reality. This is a constant refrain throughout the *Lectures*. 'It [the spirit] generates out of itself works of fine art as the first reconciling middle term between pure thought and what is merely external, sensuous, transient, between nature and finite reality and the infinite freedom of conceptual thinking.'[18] As a *reconciling* term, art is no longer strictly acting in the manner of the *Aufhebung* as such, as it had done in the *Phenomenology*, where it allows shape to raise itself, through creative activity, into a higher form of consciousness, and eventually into self-consciousness. In the *Lectures* art, as fine art, does not enfold the two sides of thought and sensuousness into a higher form, one that both cancels the two sides while at the same time allows them to continue in a unity of transformation. It stands in the middle, as a mediating modality: 'It is *not yet* pure thought, but, despite its sensuousness, is *no longer* a purely material existent either'.[19] It is a kind of halfway house. But as we have seen from the *Science of Logic*, something cannot reside in the *not yet* and *no longer* simultaneously without some force of negation present to keep it (paradoxically) in that active state. Thus, early in the Introduction to the *Lectures*, Hegel provides us with a significant passage concerning the relation between thought and thinking (the *Begriff*) on the one hand and the sense and feeling pertaining to the concrete work of art on the other. It is worth quoting in its length:

> Now art and works of art, by springing from [*entsprungen*] and being created by [*erzeugen*] the spirit, are themselves of a spiritual kind, even if their presentation assumes an appearance of sensuousness and pervades the sensuous with spirit. In this respect art already lies nearer to the spirit and its thinking than purely external spiritless nature does. In the products of art, the spirit has to do solely with its own. And even if works of art are not thought or the Concept [*Begriff*], but a development of the Concept out of itself, a shift of the Concept from its own ground to that of sense, still the power of the thinking spirit lies in being able not only to grasp itself in its proper form as thinking, but to know itself again as much when it has surrendered its proper form to feeling and sense, to comprehend itself in its opposite, because it changes into thoughts what has been estranged and so reverts to itself. And in this preoccupation with its opposite the thinking spirit is not false to itself at all as if it were forgetting and abandoning itself thereby, nor is it so powerless as to be unable to grasp what is different from itself; on the contrary it comprehends both itself and its opposite. For the Concept is the universal which maintains itself in its particularizations, overreaches itself and its opposite, and so it is also the power and activity of cancelling [*aufzuheben*] again the estrangement in which it gets involved. Thus the work of art too, in which thought expresses itself, belongs to the sphere of conceptual thinking, and the spirit, by subjecting it to philosophic treatment, is thereby merely satisfying the need of the spirit's inmost nature.[20]

We can first note here that works of art spring from or are generated by the spirit, as *Geist*. Therefore they have, by their innermost nature, a spiritual origin and spiritual constitution. This much we know already from the *Phenomenology*. Their external nature already exceeds that found in the natural world; their human creative element takes them above pure nature, and links them with the divine (as *Geist*). Thus when Hegel next says that, in the products of art, the spirit 'has to do solely with its own' (*nur mit dem Seinigen zu tun*), he once again demotes the mimetic tradition, and equates art fully with the activity of human originating that involves us, necessarily, in the sphere of the spiritual. In this sense he is in concord with many of his Romantic contemporaries. But it is not simply the originating of a sensuous or concrete work of art – the aesthetic object – that Hegel has in mind here as the 'product'. What is *mit dem Seinigen*, 'with its own', is also the inherent power to self-originate that marks the spirit *qua* spirit. Spirit begets itself, just as, earlier, thought begets itself. Art, then, is not about originating something external to itself, but about *originating itself*. Again, not as an 'aesthetic object', nor as the self-contained work of art whose only interest is its own internal (material) subjectivity. Art is rather the expression of originating, which goes to the very core of spirit and its logic. This takes us back to the self-originating power we saw inherent within the dynamics of the *Science of Logic*. Thus in the next line, Hegel says the work of art, even if it is not identical to thought or the *Begriff*, allows the *Begriff* to develop, to go forth from itself, to come into the light of day from the darkness of its interior, and thus to know itself as much as the outside creature as the inside creature. The work of art is not about the 'outside' as such, the material product in its finitude, nor even about the 'inside' as such, the *Begriff* that resides in thought. Rather, it is about its *issuing forth*, its passage from the interior to the exterior, which is *Begriff*'s very constitution, a movement that becomes the development of its own resources.[21]

Now if we look more closely at this development, we find it is constituted by a power that goes two ways – an 'amphibious' power, we might say, to use Hegel's later image.[22] For 'the power of the thinking spirit' lies not only in its self-conscious awareness of its proper form *in thinking*, but also as much in the self-conscious awareness of itself *in its opposite*, in its 'surrender' (*Entäusserung*) of its thinking form to feeling and sense. Its power re-domesticates what has been 'estranged' (*Entfremdete*). The thinking spirit remains then 'preoccupied' with its opposite. Not only that, in this very preoccupation it 'comprehends' both itself and its opposite. But it does so only by 'the power and activity' of *aufheben*, the sublating of its estrangement with, and within, itself.

Hopefully, we can begin now to see why Hegel's conception of art is inextricably bound up with his entire philosophical project. Art does not imitate nature outside us. Nor does art merely show us ourselves in a more concentrated external form. Nor does art merely provide us with new mythologies with which to broaden our imaginative capacities, or even our moral capabilities (Kant, Schiller). Art shows our preoccupation with our opposite, and enacts a certain overcoming of the estrangement that such internal opposition entails. So it belongs to 'the sphere of conceptual thinking', but a sphere that is always and already driven by its own internal negations. If art is called to make the external correspond to the *Begriff*, as Hegel later summarises,[23] then it can only do so by enlisting the operations of negation to overcome its own internal estrangements.

NEGATION'S BEAUTY – SCHILLER

To see how these operations are developed more closely in Part One of the *Lectures*, let us return to the question of beauty we touched on above. Much has been written of Hegel's concept of beauty in relation to art,[24] but little has been said of beauty's own negative comportment. To understand this comportment, let us, once again, return to that important source for Hegel's understanding of the art of negation, Schiller's *On the Aesthetic Education of Man*.

We recall from our discussion in Chapter 1 Schiller's three impulses or drives that help overcome the division between our subjective selves and the objective world around us: the sensuous impulse (*sinnlicher Trieb*), the formal impulse (*Formtrieb*), and the play impulse (*Spieltrieb*). We saw that the latter combines the former two (life and form) in providing what Schiller had called a 'living shape' (*lebende Gestalt*). He then tells us that this latter shape 'serves to denote all aesthetic qualities of phenomena and – in a word – what we call *Beauty* [*Schönheit*] in the widest sense of the term'.[25] Certainly, he has borrowed from Kant's philosophical ennobling of the beautiful in his *Critique of Judgment*, particularly where the beautiful comes into its own in relation only to itself and not to any particular end or interest outside itself, and where beauty can lead us, through the freedom of this disinterestedness, to a higher moral ground. For Schiller, we can *play* with beauty, as we do in art, but this play is more than frivolous gaming or *divertimento*. It reconciles the two competing sides of our nature, the material and the formal: 'Through Beauty the sensuous man is led to form and to thought; through Beauty the spiritual man is brought back to matter and restored to the world of sense.'[26]

Now this reconciliation is an advance on Kant because, rather than making reflective judgement the standard, at the expense of sensation, it keeps the two sides in a productive tension.[27] The *Trieb* already indicates to us the generative nature of the reconciling power. To properly examine this tension, and to see if, in its reconciliation, the *Trieb* actually resolves the tension, or does away with it altogether, requires a certain conceptualisation of Beauty (Sixteenth Letter). In conceptualising Beauty we find that the two sides are made that much bolder, and appear now in an antagonism as irreconcilable opposites, two radical enemies who can never become one. Yet by means of the impulse that is play, says Schiller in the Eighteenth Letter, Beauty will be able to overcome this opposition, and combine the two sides. And he will want this combination to be comprehensive, leaving no residual elements on either side. But in describing this process, Schiller draws upon the language of sublation: 'Beauty *combines* those two opposite conditions, and thus removes the opposition. But since both conditions remain eternally opposed to one another, they can only be combined by *cancellation* [so sind sie nicht anders zu verbinden, als indem sie *aufgehoben werden*].'[28] Now previous philosophers only ever tried to deal with Beauty through a protection *against* cancellation or *Aufhebung*: those bound to feelings fought against the cancellation of Beauty 'as an operative power [als wirkende Kraft aufzuheben]', while those bound to logic fought against the cancellation of Beauty 'as a concept [als Begriff aufzuheben]'.[29] Yet actual Beauty must survive not against *aufheben* but within it, in order that both sides are fully assumed into a single unity, even if, in that unity, 'both those conditions completely disappear.'[30]

Now Schiller has not yet understood the *Aufhebung* as Hegel will understand it later, that is, as a cancellation towards, not mutual disappearance, but mutual preservation. But he is inching towards that understanding. In the subsequent letter, the Nineteenth, he comes at the question of Beauty by way of the question of determinability (*Bestimmbarkeit*) or determination (*Bestimmung*). Following Kantianism, Schiller sees the path to determination through two faculties: first, the senses (*Sinne*); second, the imaginative powers of conception (*Vorstellung*). The first is infinite in its possibility, for the senses in their passivity have an endless array of impressions before them, and without some limiting force the world would remain indeterminate. The second is just such a force, an active, operative power (*wirkende Kraft*) that *sets limits* in order to determine a reality that can be grasped. Yet just as Hegel will elaborate in the later *Science of Logic*, Schiller sees an inherent contradiction in the movement from the first to the second. For though limiting (*begrenzen*) creates a reality we can behold and understand, it does so by exclusion of all other possibilities in their infinity. That is, it implements an active negation: 'So we arrive at reality only through limitation, at the *positive*, or actually established, only through *negation* [*Negation*] or exclusion, at determination only through the surrender [*Aufhebung*] of our free determinability.'[31] And this contradiction can be seen as an activity of thinking itself, for only by an 'absolute act of the mind [*absolute Tathandlung des Geistes*]' can the negation become related to something positive, and this, says Schiller, 'is called judging or thinking, and its result is *thought*.'[32]

We can see the incipience of Hegel's thought here, beyond Kant. Thinking generates by way of an *active* negation – hence Schiller's use of *Vorstellung*, and not *Begriff* as Kant had used, to describe the originating powers of determination. (*Vorstellung* carries the sense of performing, the movement of placing or positioning (*stellen*); while *Begriff* carries the sense of arresting something through seizure or grasping (*greifen*)). Hegel will of course transmit Schiller's *Vorstellung* into his own appropriation of Kant's *Begriff*, and the *absolute Tathandlung des Geistes* will take on new meaning through the *Phenomenology* and the *Greater Logic*. But here in this passage of Schiller's, Beauty now returns, not as something that can somehow insert itself between sense and thought, to fill up the gap, as it were, but as something that helps maintain, or better, that provides the mode for, the freedom of thought to express itself *according to its own laws*. Committed to Kantian freedom as he is, Schiller wants to preserve the autonomy of the thinking mind in its fundamental freedom to think and produce thought from within its own internal jurisdiction (otherwise it is not truly free). If the sensuous impulse and the formal impulse are in opposition to one another, it is only in the freedom of the play impulse (and not in reflection upon one side or the other) that their necessities can be overcome. In this way, and only this way, Beauty 'can become a means of leading Man from matter to form, from perception to principles, from a limited to an absolute existence.'[33]

But Schiller acknowledges a problem. For if Beauty, in its freedom, establishes and maintains the mind as self-productive or self-determining, then the mind too falls prey to its internal contradictions. For if, in its determining capacity, it must 'surrender [*Aufhebung*]' its 'free determinability', or impose limits from within, then it is neither wholly free nor wholly active, since it gives itself over to the limit and surrenders its activity to a passivity. Or we could say, the active power of the mind is not involved in a creation *ex nihilo*, since sensations are always something given to us. The active power is always in some sense *via negativa*, since it must limit the givens to make sense of them. As Schiller puts it, 'it is not enough for something to begin which did not previously exist; something must first cease which previously did exist.'[34] And it is not just the sensations themselves that must cease; it is also the *determination*. Schiller elaborates in a significant passage:

> In order, therefore to exchange passivity for self-dependence, an inactive determination for an active one, he [Man] must be momentarily free from all determination and pass through a condition of mere determinability. Consequently, he must in a certain fashion return to that negative condition of sheer indeterminacy in which he existed before anything at all made an impression upon his sense. But that condition was completely devoid of content, and it is now a question of reconciling an equal indeterminacy and an equally unlimited determinacy with the greatest possible degree of content, since something positive is to result directly from this condition. The determination which he received by means of sensation must therefore be preserved, because he must not lose hold of reality; but at the same time it must, insofar as it is a limitation, be removed, because an unlimited determinacy is to make its appearance. His task is therefore to annihilate and at the

same time to preserve [*zugleich zu vernichten und beizubehalten*] the determination of his condition, a thing which can be done in only one way – by opposing that determination with another.³⁵

To be active both ways, both in sensation and in reason, and simultaneously, also means a mutual destruction of both modes: 'they are mutually destroying their determining power and through their opposition *producing negation*.'³⁶ And the condition which allows us this simultaneous activity, this production of negation, is precisely what Schiller then calls *the aesthetic*. The fact that it is *productive*, even of negation, is what we must hold on to here.

NEGATION'S BEAUTY – HEGEL

To see how Hegel adapts Schiller, and extrapolates what he has said in the Introduction of the *Aesthetics*, but now in the more sustained context of Beauty, let us turn to Part I, 'The Idea of Artistic Beauty, or the Ideal'.³⁷ In the opening section, Hegel wants to position art in relation to the finite world and to religion and philosophy. To do so, he returns to three basic terms: the Concept (*Begriff*), the Idea (*Idee*) and the Spirit (*Geist*). These terms are not identical; but they are intimately related, one to the other.

From the *Phenomenology*, we know that *Geist* is self-consciousness in absolute knowledge of itself, Spirit knowing itself as Spirit. By the end of the *Phenomenology*, this *Geist* stands in direct relation to the *Begriff*, for in its self-conscious knowing it 'shapes' itself through the *Begriff*, to become 'thought which begets itself as thought'. The *Begriff*, then, is the pure element, as pure activity, of *Geist*'s existence.³⁸

In the *Science of Logic*, the *Idee* is defined as the unity of the *Begriff* with objectivity or reality, as opposed to merely the *Begriff* in itself. The *Idee* is, we might say, the *Begriff* in-and-for-itself.³⁹ As such, it carries both a subjectivity and an objectivity simultaneously. Now Hegel says that these two sides, as opposites, relate to one another, as always, by way of negation. The former, the subjectivity of *Begriff*, is what he calls the *urge* (*Trieb*) to sublate (*aufheben*) the difference of the two opposing sides, and remain 'in-itself'; the latter, the objectivity of *Begriff*, is indifferent to any such prioritising, and is happy to simply subsist 'for itself' in its positedness, and thus make any in-and-for-itself 'null'. The *Idee* is the *identity* of these two sides. But in order to make them identical, it must nullify the urge of the first towards subjectivity and nullify the nullification of the second towards objectivity. It must, that is, forever negate its own internal opposition, which, as the negation of negation, is itself in perpetual opposition. This is why Hegel says 'the Idea possesses within itself the *most stubborn opposition*'.⁴⁰

This stubborn opposition returns now in relation to Beauty. In the Introduction to the *Aesthetics*, we are told that art has the task of presenting the *Idee* not in the form of thought, as it is customarily understood and received, but in the form of the sensuous shape. Following Schiller, art must therefore unify two differing realms, so that 'the loftiness and excellence of art in attaining a reality adequate to its Concept [*Begriff*] will depend on the degree of inwardness and unity in which Idea [*Idee*] and shape [*Gestalt*] appear fused into one.'⁴¹ But the *Idee*, as fused, cannot be just any idea – it must be a universal idea, and the universal idea of Beauty itself.

In the opening of Part I, this idea, as *Idee*, is taken further. One of the inherent features of Beauty is that it must always be connected to some form of objectivity. The beautiful must be something we can set our eyes upon, even if that sight transports us to some other imperceptible realm. It thus naturally lends itself to the *Idee*, the *Begriff* in unity with objectivity or reality. Hegel affirms that Beauty cannot ever be merely an abstraction; it is rather an 'inherently concrete absolute Concept', the *Idee* with all its stubborn oppositions. But if it is going to be a universal *Idee*, it must be absolute, wholly adequate to itself, not contingent on something external to it. And if it is going to be an absolute *Idee*, it must also be, Hegel says, an 'absolute spirit',⁴² for only in *Geist* can the total adequation of the inward and outward be held and drawn into some kind of manifestation. Art then is brought back to the context of the *Phenomenology*, for from art the *Geist* makes its trajectory through religion and philosophy as the inwardisation of the outward and the outwardisation of the inward. 'The realm of fine art is the realm of *absolute spirit*.'⁴³

Beauty naturally accompanies this trajectory. Traditionally, Beauty has always been conceived in the coupling of nature and spirit. The aesthetic realm has retained its august status precisely because it allows nature to penetrate the world of spirit. But we know that Hegel's *Geist* is not so passive. It does not simply allow itself to be penetrated by nature, nor even to penetrate nature in turn, under some invitation inherent within the aesthetic process. For as we know already from the passage we cited earlier in the Introduction in relation to art, the *Geist* is generative in its nature. It thus *produces* nature, and produces it out of *absolute activity*. Nature is *Geist*'s product, or better, its creation. But this creation does not stand outside *Geist*; it *is Geist* in its innermost being, nature and spirit unified. And this is *Geist* as Beauty, the Beauty we almost always first encounter as the beauty of nature. This is why in Part I, after dealing with Beauty at the conceptual level, Hegel pursues the question not first in relation to art but in relation to nature.

But even at the level of nature, this activity of creation leads to *Geist*'s own internal diremption. For creation, as *Geist*'s ownmost activity, is self-creation, and as we know from above (thought that begets itself), self-creation requires self-differentiating: what is produced or created is something other, but an other *within*. And so to say nature is the other of *Geist* is to say that it is *Geist*'s other within, nature born of the spirit. Nature therefore has the *Geist* implicitly within it, just as the *Geist* has nature implicitly within it. By extension, we can also say that Nature is implicit in the *Idee*, and the *Idee* in nature. But it is only so by negating itself through productive negation. *Geist* is thus both 'ideality and negativity'. Hegel elaborates: 'spirit particularizes itself within and negates itself, yet this particularization and negation of itself, as having been brought about *by itself*, it nevertheless cancels, and instead of having a limitation and restriction therein it binds itself together with its opposite in free universality. This ideality and infinite negativity constitutes the profound concept of the *subjectivity* of spirit.'[44]

If ideality and negativity constitute *Geist*'s subjectivity, we know from the negations of the *Logic* that its objectivity is closely correlated. Indeed its objectivity is the finitude and concreteness that *Geist* produces within itself as its own self-differentiation, as nature. But in order to be objective within, it must know this finitude and concreteness as its very own: it must sublate its own subjectivity knowingly, and in turn must also negate that negation knowingly, knowing its subjectivity as objective and its objectivity as subjective, and, ultimately, its negativity as constitutive for both sides. It cannot stay on one side only, for then it remains restricted, deficient, in a state of 'unrest', 'grief', 'something *negative*'. It has to sublate this negativity, even if the sublation itself owes its very existence and power to this negativity.

There are thus two activities that coincide here: the activity of knowing, and the activity of sublation. In the first, the knowing is an absolute knowing – knowing wholly adequate to itself – and as Absolute it becomes the very object of *Geist*: 'the Absolute as spirit and self-knowledge.'[45] In the second, the sublation operates with absolute negativity – negation wholly adequate to itself – and as such is always *producing* (its own) negation. Now exactly at this coincidence, where the production of negation becomes identical to the production of knowing, where knowing is born from sublation's oscillations between negating and reconciling, exactly here, says Hegel, is where 'we have to begin in the philosophy of art'.[46]

But let us be careful. It is not simply that art is the product that somehow arises out of these oscillations, as an *aesthetic* product. The aesthetic product is nowhere yet on Hegel's radar. The beginning of art is its philosophical beginning, art as *Begriff*, *Idee* and *Geist*. But it is precisely here where art properly *begins*, not as the product, not even as the product of knowing or sublation, but as the oscillating power itself, the force of that to-ing and fro-ing between opposites. This is why, in beginning with the *Begriff*, *Idee* and *Geist* of art, one is also beginning with the *Begriff*, *Idee* and *Geist* in and of themselves. 'The realm of fine art is the realm of *absolute spirit*' means nothing other than this.

In sum, Beauty for Hegel, as for Schiller, is far more than an aesthetic quality. Beauty, as the passage between sensuous experience and formal thought, is also the overcoming of that passage, the sublation of the opposites. The *concept* of Beauty, the *idea* of Beauty, the *spirit* of Beauty – each of these is grounded in an infinite negativity: 'not a negation of something else, but self-determination in which it remains purely and simply a self-relating affirmative unity.'[47] This is Hegel's reinterpretation of Schiller's Beauty, now in the maturity of a full *Aufhebung*.

It is also Hegel's reinterpretation of Schiller's 'living shape'. The activity of sublation is the very stuff of life. So Hegel expounds in an extraordinary passage:

> To go through this process of opposition, contradiction, and the resolution of the contradiction is the higher privilege of living beings; what from the beginning is and remains *only* affirmative is and remains without life. Life proceeds to negation and its grief, and it only becomes affirmative in its own eyes by obliterating [*Tilgung*] the opposition and contradiction.[48]

Art is living because art proceeds to negation. But it is only possible for it to proceed *to* negation because it first proceeds *from* negation. Art is this circularity, this cogenitive power to oscillate *to* and *from*. It is yet more: a generative power that, in its absolute capacity, creates the *to* and the *from*, the pre-positions of *Geist*. This creation is ultimately what Hegel wants to call the beautiful.

It is only now we are in a position to look at art in its more customary sense: the products of aesthetic creation in their various forms and individualities. So far, we have not progressed beyond the foundations of art as *Begriff*, *Idee* and *Geist*. There is so much yet to come in the name of general classifications, historical developments, systematic genres and specific examples – the *arts*, in all their finery. But what we have just uncovered above is the very essence of that plurality, the very possibility for any aesthetics whatsoever. And this 'essence' is a power of negation that estranges art from the aesthetic comforts of its taxonomies and formalities. It is not that negation restrains the development of art towards aesthetic reflection or aesthetic analysis. Clearly, Hegel is under no such restraint himself, as his *Lectures* proceed towards the voluminous detail of art in all its historical formations and categorical particularities. But in that development negation is never not working. And we might say, despite itself. For the Beauty of art is ostensibly manifested in certain world-historical states and in specific forms of art, but only through an active development in which *Geist* at the same time is manifesting its absolute nature. Art, we repeat, is never disunified by the *stasis* of reflection. Art is always in the *kinesis* of unfolding itself. This is the 'system' that governs the movement of the *Lectures*: art advancing towards its absolute unity in *Geist*. But this advance is predicated on the repeated negations at its innermost core. To be *life*, to be *living* shape, to be animated by the organicism of nature and spirit's inseparability, requires the surrender, the estrangement, the contradiction, the oscillation of opposing movement. Beauty is Beauty precisely because it impels a unification through the movement of negation. Hegel might be able to delineate the features of an individual living shape, but only with the purpose of showing how that shape gets further shaped beyond itself, how its own individuality is surpassed towards the absolute. This is not an absolute that then leaves behind all traces of individuality, as in some Kierkegaardian complaint, but an absolute whose own features – universality, infinity, purity, truth, adequation without remainder – are marked by an absolute negativity, leaving its own core dirempted to the same degree it is unified. This 'system' is expressed conventionally, and conveniently, by the unity. But it is generated, invisibly, by the kinetic power that is negation. To make this power visible is the task of a philosophy that dares to unify itself, diremptively, with art. To read Hegel's *Lectures* as if they were principally *an* aesthetics, the explication of art *qua* art, misses the invisible forces underlying them. Rather, the 'philosophy of art' is to be read cogenitively, art implicit in philosophy, philosophy implicit in art. To read it any other way is to prioritise one side of the opposition at the expense of the other, the nature of art or the spirit of art. But the art of negation nullifies any one priority. Even, contradictorily, its own. For that contradiction is precisely its ownmost affirmation.

Now from here, we could spend much time, as others have done, bringing to the fore the various lineaments of art in what Hegel calls its 'ideal' form: the *Begriff*, *Idee* and *Geist* now in their determinate and individual actualisation as beautiful works of art. This would entail broader discussions of art in its historical phases – Classical and Romantic – and of their corresponding forms, sub-forms, genres, types and styles, all of which occupy the remainder of the *Lectures*. But it is not our purpose to trace out the contours of these parts or the particular ligaments and sinews that hold them together. It is our purpose to show their motive force, their *kinesis*, that which animates them into a living unity of material and spirit, an absolute corpus. To do this, we must surpass the analysis of art as aesthetic products. This surpassing returns us to the beginning, to art as *a priori*, the *a priori* of a generative function. But this function is also the function of a surpassing, an absolute surpassing, an *a posteriori* in perpetual motion. To see this surpassing at its most comprehensive, let us conclude this chapter on *Aesthetics* by returning to the much worked question of the 'end' of art.

ART'S END

The supposed 'end of art' thesis has become to the *Aesthetics* what the master–slave dialectic has become to the *Phenomenology*: a pericope with a life of its own. Those readers who do not endure all the intricacies of each respective text come away at least with these two moments, even if those moments harbour their own intricacies well below the surface of their popular interpretations. We cannot question the attention given to them, nor should we puzzle at all the various interpretations they have generated, for they are worthy passages, intriguing each in their way. What we can offer, in the case of art's apparent end, is an interpretation that now ties that end to the surpassing negations of the foregoing discussion.[49]

Any thesis of art's end or death has been drawn, almost invariably, from the opening pages of the *Aesthetics*, to which we alluded earlier in the Introduction. There, we remember, Hegel says: 'In all these respects, art, considered in its highest vocation, is and remains for us a thing of the past.'[50] And the 'respects', we saw, pertained to the positioning of art within the same sphere as religion and philosophy, which, we now know, means in respect to *Geist*, or to Spirit in its absolute self-knowledge and self-fulfilment. Now since that *Geist* is in progress, it must fulfil its own actualisation according to the various stages of its historical manifestation. If, in the Classical era, the artist and the expression of the divine were in total accord, so that the one was not in any way the representation of the other, but in fact the actual generation or *poesis* of the other, and one did not need to reflect on art to penetrate through to the divine, because art *was* the divine, in all its immediacy, and vice versa, then that stage has now been surpassed. If, in the Middle Ages, the christologies of incarnation vaunted the sacred work of art to the point of veneration and, *in extremis*, worship, then that stage too has been surpassed. Greece's theogonies/mythologies and Christendom's sacramentalism have now given way to a more inward disposition, so that the manifestation of *Geist*'s actualisation must now be sought in a *reflective* mode, and art, if it is to maintain its 'highest vocation' of effecting the truth and life of *Geist*, must too give way to higher, more reflective modes, that is, to religion and philosophy. Thus, for us in this more reflective stage, art becomes a thing of the past.

But this passage is not the only instance of the 'end of art' in the *Aesthetics*. In the opening section of Part I, Hegel returns to this 'end', but now in a much more developed manner. Part I, we have just seen, elaborates the idea of Beauty, where Beauty becomes both the passage between sensuous experience and formal thought, and the overcoming of that passage, in a sublation of opposites. That sublation brings Beauty into the realm of the Absolute, as the truth of *Geist* developed through both *Begriff* and *Idee*. If therefore truth exists as the content of these three fully integrated correlatives, *Geist*, *Begriff* and *Idee*, so too it exists as the content of three fully integrated forms: art, religion and philosophy. Thus, these 'three realms of absolute spirit differ only in the *forms* in which they bring home to consciousness their object, the Absolute'.[51]

The forms in question here are more than indifferent modalities. They are tied inextricably to the activities of thought *as* thought. Thus, Hegel differentiates the three in relation to epistemological relations: art is 'sensuous knowing', religion is 'pictorial knowing' and philosophy is 'the free thinking', each in respect of absolute spirit.[52] If we keep in mind our epistemological axiom, *thought that begets itself*, each of these forms then is involved in the activity of self-begetting. It is for this reason Hegel says here that art, at its highest, is never constituted by things outside itself, either in the manner of external interest or of utility. Art's 'subject' is itself alone, which is to say, the truth of *Geist* in its sensuous manifestation. So Hegel says, with a view back to the ancient Greek unity of art and the divine, 'This is the original true standing of art as the first and immediate satisfaction of absolute spirit.'[53]

But if this 'original true standing' constitutes a 'before', a stage that has long been surpassed, then art also has its 'after'. This is *not* to say – and many commentators have pointed this out – that it has its 'end' or its 'death'. It is to say rather that art carries both sides within itself: an original stage the perfection of which it can no longer incarnate, and a subsequent stage to which it must now accede. So Hegel repeats the earlier 'a thing of the past' with new articulation: 'For us art counts no longer as the highest mode in which truth fashions an existence for itself.' Art, instead, 'points beyond itself'.[54] It does not expire; it is rather translated into an 'after' stage more true to the *Geist* in its progressive unfolding. 'Thus the "*after*" of art consists in the fact that there dwells in the spirit the need to satisfy itself solely in its own inner self as the true form for the truth to take.' It becomes a question of need: since *Geist* can no longer perfect (make absolute) its truth in sensuous shape alone, it must yield to 'higher' forms – first religion, then philosophy. Hegel expresses here the hope that art might one day rise to the occasion again, but he acknowledges it will not be in its present form: 'We may well hope that art will always rise higher and come to perfection, but the form of art has ceased to be the supreme need of the spirit... we bow the knee no longer.'[55] For all this, however, art does not disappear. Its translation is towards a unity with philosophy. It surpasses itself in order that its sensuousness and objectivity might be taken up by the *thinking* that is philosophy. Yet that thinking, as we have been at pains to show, is also a thinking of its own sensuousness and objectivity, even if from a subjectively reflective position. So its translation is in fact a sublation, and art is never not at work in the activity of philosophical thought, as it is never not in play in a religion that too must accede its own form.

Let us now carry the implications of this 'never not' to its furthest reach, perhaps even beyond the comforts of Hegel's own system. If yielding up its own form is inherent in the process of art and religion's formation, this yield goes both ways: it produces itself while at the same time it surrenders itself. If this yield is the yield of negation, as our extrapolations above have shown, then philosophy does not escape such negative activity. For thinking, as the cogenitive thinking of philosophy, is itself precisely the power to generate itself out of its own yielding up: production and surrender in absolute mutuality. Only in a restricted historical sense (that is, chronological or diachronic) could we say art and religion in their formation have surpassed themselves while philosophy, as the culmination and consummation of *Geist*, has remained untouched. But we know this is definitively *not* how Hegel reads history. History is not differentiated from the formations of thinking; history is implicated in thinking's very origination. And this origination, we have shown, has its wellspring in the power that is negation. If art's translation to a higher form is in fact a sublation, then inherent in this process is all that we have been arguing about the nature of *Aufhebung* – its origination in and as negation. Philosophy is not somehow exempt from this *Aufhebung*, by virtue of some historical or chronological privilege. Philosophy – thinking – *is* this *Aufhebung*. And its origin is art, the art of negation at its very core.

If therefore art is surpassed in the scheme of unfolding spirit, so too is religion, and so too is philosophy. The truth of each of their forms, the content identical to them all, is self-generated out of the same power that is negation's living potency. If there is an 'end', an 'after', a 'pastness' to art as it emerges from the *Aesthetics*, so also is there an 'end', an 'after', a 'pastness' to religion and to philosophy, both *a priori* and *a posteriori*. Had we stopped at the *Science of Logic*, this apposition might not seem so radical. Read through the subsequent writings, it overturns any successive schema within the system, whereby art must give way to religion, which in turn must give way to philosophy, the apex of human endeavour. Each form, each stage, each mode of thinking is never not without the art of negation at its very origin. And this is the disruptive nature of the process – negation overturns even the stabilities of logic, as a science or as a philosophical system. Likewise, it overturns aesthetics, as the reflective engagement with aesthetic production. The *philosophy of art*, under negation's regime, is always cogenitive, and therefore always unstable as nominative thinking. The philosophy of art, under Hegel's negation, is an active making anew. We might even say it is a returning anew: returning philosophy to its ownmost originations and disruptions, at the core of which is the art of self-begetting, art *as* self-begetting. This return carries the *Aesthetics*, even after all its expatiation and elaborations, back to the *Phenomenology* and the *Greater Logic*. But more, it carries negation into the very heart and soul of any such system one might construct out them. It is this 'return' that best marks the 'return of Hegel' in our contemporary philosophy and our contemporary world. And it is to this world we now turn.

PART II

THE NEGATION OF HEGEL

CHAPTER 4

The Returning of Hegel and Negation: Sartre and Hyppolite

When we turn to the latter-day conditions of our world, one thing is salient: negation is everywhere present. We do not mean this merely in the sense of a negative attitude, which permeates a society under the constant pressure of political, economic, military and/or cultural turmoil. To be sure, rampant media access and news coverage do not let us avoid the instabilities, conflicts, injustices and hypocrisies generated daily on a global level. And we are more prone to pessimism and cynicism under the exposure of these realities (and the realities of this exposure) than we are to optimism and hope. 'Absolute negativity is in plain sight and has ceased to surprise anyone', Adorno said, even in the 1960s.[1] But negation makes itself present at a much deeper level as well: we have become accustomed to entertaining the negative in both our natural cosmologies and our existential or philosophical ruminations. Empirically, we have accepted the central position that the negative must play in the scientific paradigms of our world, whether in the mathematical concept of zero and negative numbers with which we set about our modern computations, or in the astrophysical hypotheses about the structure, and indeed origin, of our modern universe. Rationally, by the same token, we have been trying to come to terms with a growing and pervading sense of nihilism. Much philosophical energy has been spent in counterforce to the various nihilisms that have mounted a charge since the *fin de siècle* of the nineteenth century, whether under the high-set flag of Nietzsche, whose nihilisms (and he understood there to be many) continue to be misunderstood, or under the flag of his intellectual successors, whose allegiances extend, like

Nietzsche's, well beyond the discipline of philosophy alone. But much energy has also been spent, particularly during the twentieth century, reworking the negative, either to rehabilitate nihilism as a viable way of thinking about the world and its conditions, or to employ the negative towards some theoretically critical end, in the heritage of *Ideologiekritik* or, more recently, of a 'deconstruction'.[2]

This pervasive acceptance therefore cannot, despite the temptation, and sometimes insistence from certain quarters, be called a malaise. One might invoke, in the name of a now canonical reaction to Hegel, a 'sickness unto death', and blame the perpetuation of negativity on an entrenched despair that has found its expression in the depths of existentialism, whether in its Christian or in its later atheistic versions. But a self unreconciled to itself, either because of sin or because of its ineluctable mortal condition, is the same self that has acknowledged and oftentimes celebrated the vacuums, the abysses, the black holes, the unconscious drives, the nothingness of the 'real', and the silent lacunae of language and meaning that make up or govern our universe, our world and our selves of late modernity. And if there are zero-sum economics, zero-sum politics, zero-sum military strategies that press us into disconsolation and hopelessness, there are also the zero-precision technologies and the zero-effecting virtualities that keep us inured to the global aggressions, the social hostilities and the individual *ennui* that continually threaten to overwhelm us. Nor has our encounter with the negative been shown to supress or to vitiate our creative imagination. On the contrary, it has broadened our room for creative manoeuvre. Defying any 'end of art' thesis, the arts have not suffered under negativity, but in fact have opened themselves up to new perspectives, to new, we might say, vanishing points.

The question now is how and where we trace Hegel's re-emergence in all of this, or where we locate his 'return'. We know that Hegel has never gone away. But where he has returned precisely *in relation to* the negative, and particularly to its surging nature since the latter half of the twentieth century, is what interests us now. Let us start at what might, for some, seem an obvious place, even if it requires a step back to the middle of World War II: Jean-Paul Sartre's *L'Être et néant*, or *Being and Nothingness*, first published in 1943.

SARTRE: *BEING AND NOTHINGNESS*

There is much we could say about the entry of this text into a period when world negativity was at an unprecedented height, and the extent of European and ethnic annihilation was just coming to light. Certainly in France, the question of nothingness was all too palpable, and if the French public might have understood this question differently than Sartre, whose 'phenomenological ontology' of the subtitle would have been far from the average grasp, still, they would have been all too aware, after two devastating global conflicts, that in the reality of being there resides a 'not', perhaps even now as a necessary condition, a non-being that is 'a perpetual presence in us and outside of us', a nothingness that *haunts* humanity.[3] When the English translation emerged thirteen years later in 1956, the war was well over, but the assessment of damage was ongoing, and reached to the very core of the existential: a re-evaluation of what it means to exist as a human, and how that existence must now include a negativity rendered ineradicable by the recent destructions of two world wars, and by the spectre of the Holocaust that would loom large over Western conscience for decades to come. That *Being and Nothingness* became a 'popular' book among the growing disillusioned of the late 1950s and the '60s indicates how pervasively the sense of negation had filtered through to the cultural 'imaginary', to borrow terminology originated by another with similar foresight, Lacan.

Sartre's main interlocutors in *Being and Nothingness* come out of the phenomenological tradition – Husserl and Heidegger. But Hegel is not not present, to double a negative in a manner germane to the text. Certainly, by 1943, Sartre had read his Hegel; but he had done so in a limited fashion, and often through the lens of others, especially Alexandre Kojève.[4] The central text on negation, the *Science of Logic*, seems largely and curiously absent from an 'essay' so devoted to the concept of nothingness. And this absence leads to a misunderstanding of Hegel's negation as it is briefly treated in Sartre's first chapter, 'The Origin of Negation'.

In Sartre's ontology, being, as Being, can only be what 'is' in pure plenitude and positivity. He states this very early on: being is what it is, that is to say a fullness of its existence that cannot admit anything other than its own 'isness'.[5] Here Sartre adopts Hegel's language of 'in-itself' (*an sich*, or in French, *en-soi*). Being-in-itself is this being that is what it is by virtue of being wholly and positively existent. It cannot, in any way, be anything other. But such existence, as an existent, is 'static' – it involves no movement as such, since any movement is arrested in the plenitude of its being; it simply *is*. Only when it 'moves' to consciousness does it leave its resting place of plenitude. It then arrives at the 'for-itself' (*pour-soi*, or Hegel's *für sich*), the point at which negation enters its field. But that arrival too is static, for it does not arrive by means of a perpetual motion between two points, itself and its opposite. Its arrival is already in place by means of a consciousness that admits something other than its own pure positivity: consciousness, which, in true phenomenological fashion, is always consciousness *of* something. Since this consciousness is what it means to be human, the *human* being is always for-itself, even if it has its unconscious in-itself as its base.

But this relation of in-itself and for-itself has no ontological development as such. (Sartre never functionally employs Hegel's in-and-for-itself (*an-und-für sich*). He calls it an 'impossible ideal'.)[6] And this is why, ontologically, Sartre cannot countenance any sense of becoming. He writes very pointedly at the end of the Introduction: 'Transition, becoming, anything which permits us to say that being is not yet what it will be and that it is already what it is not – all that is forbidden on principle. For being *is* the being of becoming and due to this fact it is beyond becoming.'[7]

This of course contrasts starkly with Hegel. The dialectic is such only *as movement*. In the *Phenomenology*, it is true that the 'in-itself' is the self relating wholly to itself, without ever going outside itself. But that in-itself is never static; it is always in movement to its for-itself. From the Preface onwards such movement is unrelenting, and we cannot read the *Phenomenology* without being caught in its powerful current. Moreover, we soon come to realise that this movement, as the movement of the self within itself, is nothing other than negation. Already in the Preface this is made clear. The internal distinction between the conscious 'I' and the substance that is its object is the '*negative* in general... their soul, or that which moves them.' Hegel iterates: 'That is why some of the ancients conceived of the *void* as the principle of motion, for they rightly saw the moving principle as the *negative*, though they did not as yet grasp that the negative is the self.'[8] So when later he speaks of a unity within either the self or a thing the self might perceive, Hegel calls such unity a 'moment of negation', for they can only be unified out of the disunity of their internal oppositions, where both the disunity and the overcoming of the disunity, as movement, is negation.[9] Their unity is precisely the in-and-for-itself (*an-und-für sich*).

In the *Science of Logic*, this movement takes place at the very centre of Being, in the activity of its purity, long before the being-for-itself (as an infinite relation of determinate being to itself) is reached. We have already seen how being, nothing and becoming work their triadic movement in the opening pages of the *Logic*'s first chapter. Becoming, we saw, was absolutely essential to this movement, even if it is itself sublated in order to reach the finitude of determinate being. Had Sartre read the *Science of Logic* with due care, he'd have understood that movement must *be* part of the very ontological structure of being, and that without it, one cannot *be* even in the plenitude of its 'isness'. (Hegel stays clear of the language of 'plenitude'; movement is rather the dehiscence of being.) As negation, this movement is never not operating, never not in force.[10] As we stated earlier, the 'is' is generatively copulative: it *begets*. This is something Sartre cannot accept. Possibility, he says, is part of the structure only of for-itself.[11]

Thus, when Sartre treats Hegel's nothing in the first chapter on 'The Origin of Negation', he cannot make complete sense of the dialectic that propels nothingness as negation. (It does not help that, from his own footnotes, he appears to rely almost exclusively on the *Encyclopaedia Logic*, the *Lesser Logic*, and not the *Greater Logic*.) He admits to the difficulty of allowing being to surpass its own self, and writes: 'To affirm that being is only what it is would be at least to leave being intact so far as it is its own surpassing.'[12] But to leave being 'intact' is to abjure the rupture, the diremption, the dehiscence that is its very self, and to misunderstand the surpassing, as *Aufhebung*, altogether. The point of the surpassing is that it *is*, precisely, being's 'what it is'. To be is to surpass. So when he writes, 'We see here the ambiguity of the Hegelian notion of "surpassing" which sometimes appears to be an upsurge from the inmost depths of the being considered and at other times an external movement by which this being is involved', he counts the cost of his unfamiliarity with the *Science of Logic*. For the *Aufhebung*, as surpassing, is never external. It is always the purity of an inner movement, even if it is not always an *inward* movement.

The nothing for Sartre, then, does not gain its fundamental character from Hegel. For Sartre, Hegel's logic of nothing is only a 'play on words', which in allowing being, even a being rent from determination or content, to be both 'is' and 'is not' at the same time, simply overlooks the fact that being nevertheless still 'is', and no one can cause it not to be. Thus Sartre's insistence on positivity: 'Negation can not touch the nucleus of being of Being, which is absolute plenitude and positivity.' Negation can have no 'presence' at that nucleus: 'Non-being is denied at the heart of Being.' Thus Sartre's Parmenidean colours are made clear for all to see: 'In a word, we must recall here against Hegel that being *is* and that nothingness *is not*'; or, 'we must be careful never to posit nothingness as an original abyss from which being arose.'[13]

What then *is* nothingness for Sartre, if it is not Hegel's nothing, as negation? Nothingness for Sartre only comes *after* Being, in the for-itself of consciousness. Nothingness, he says, 'is logically subsequent to it since it supposes being in order to deny it';[14] or again, 'being is prior to nothingness and establishes the ground for it... being has a logical precedence over nothingness.' As the exact reversal of Hegel's nothing as negation – 'that it is from being that nothingness derives concretely its efficacy'[15], and not the other way around – Sartre's nothing is what he calls the 'nihilation' of being in its pure positivity, which is not a co-instantaneous act but only and always a subsequent act within the for-itself of consciousness. When one is conscious *of* something, suddenly the 'other' of that something introduces ambiguity, uncertainty, indeterminateness. This disruption then redounds back onto the in-itself, which in its pure positivity finds it can no longer maintain that positivity. The 'other' is, for the in-itself, not what it is, either fully or partially. And the same goes for consciousness: '*consciousness is not what it is*' because '[c]*onsciousness of the Other is what it is not.*'[16] Sartre gives several concrete examples of this nihilation, whether in the form of a pre-arranged meeting at a café with a friend who does not show up, or a simple glance received from another across a room, which is difficult to interpret. In each case, the in-itself must give itself over to the vicissitudes and interrogations that come with being conscious of something – why? what? whence? who? and so on – and in doing so surrender its plenitude to the decompression that such interaction, through the for-itself, brings.[17] As Sartre states later, the nothing 'does not belong as an internal structure either to the thing or to the consciousness, but its being is *to-be-summoned* by the For-itself across the system of internal negations in which the in-itself is revealed in its indifference to all that is not itself.'[18]

I have written elsewhere how Sartre has a difficult time himself locating nothingness in his own schema.[19] It is neither inside, co-resident with plenitude, nor wholly outside, external as some foreign power. And yet at different times he will imply one or the other, and perhaps even both, as the previous quotation suggests. What *is* clear for Sartre is that nothingness is only a *human* phenomenon, that it pertains and operates only in regard to human beings. 'Man is the being through whom nothingness comes to the world', precisely because only the human can properly be for-itself in the consciousness of its being, and in the consciousness of others around it.[20] Another way to say this, in terms Sartre will go on to expound, is that only the human has *freedom*. Freedom inheres in possibility, and only the human being that goes out of its pure 'isness' as in-itself can encounter, and harness, the possibility that comes from the for-itself. (Will my friend show up, or has he gone to the wrong café? I could walk down to the next café to make sure. Is that glance hostile or friendly? I could have a word with the person to find out.) But such existential freedom is not the freedom of Hegel, a freedom always sublated by necessity. And neither is its corresponding existential nothingness the negation of Hegel, a negation always sublating itself. Sartre is fundamentally not dialectic in this sense.

So when Foucault called Sartre the 'last Hegelian', he did not, and could not, mean this in any sense of a dialectician.[21] Nor could he mean this in relation to a shared understanding of negation. What he was implying, rather, was that Sartre stood in a line of metaphysical thinking that goes back to Hegel, through various Hegelianisms. For ultimately, what interests Sartre is not the freedom that comes by way of the sublations of negation, but a freedom that comes by way of a synthetic unity and a totality.

In the Conclusion of *Being and Nothingness*, Sartre returns to the question of the relation between the in-itself and for-itself, and wonders how, in their apparent incommensurability, they can be seen together in the totality of being. Or how can nothingness not be within the in-itself and yet still somehow fundamentally be connected to it (as the for-itself)? He raises two significant issues that have relevance to our discussion of Hegel's return. The first is the question of self-begetting, the second is the question of synthesis. Neither accord with Hegel.

In the first, Sartre wonders if the in-itself can found itself, in the manner of what he calls *ens causa sui*, being as its own self cause. In a complicated (and hardly lucid) passage, he admits this possibility is important for the question of ontology, and even for the connection between the in-itself and for-itself. But because for something to found itself requires, for Sartre, that it be present to itself, and so have consciousness, it is caught in a contradiction, and its founding can come only through its opposite, the for-itself. Rather than embrace Hegel at this point, and accept this contradiction as vital, Sartre steps away and says not just that any such founding falls to metaphysical hypotheses (in the tradition of Kant's *as if*), but in the end it remains *impossible*.[22] That Sartre would demand self-presence for the possibility of self-founding goes to the heart of his ontology, as an ontology of positivity, and one which betrays ultimately a metaphysical commitment or allegiance.[23]

In the second, the question of synthesis, Sartre is intent on finishing his long disquisition with some understanding of how a totality can be arrived at across the divide of being's two incommensurable modes, pre-consciousness and consciousness, in-itself and for-itself. Following the tradition of Hegelianism, he employs the language of synthesis. But Sartre's 'organising' synthesis does not arise out of a sublation of the *Aufhebung*, in which both sides are cancelled and preserved in an ongoing dialectical *movement*. Rather unity is possible only by way of what we could again call a 'static' synthesis, in which the two unite only on the priority of the in-itself: 'As for the totality of the for-itself and the in-itself, this has for its characteristic the fact that the for-itself makes itself *other* in relation to the in-itself but that the in-itself is in no way other than the for-itself in its being; the in-itself purely and simply is.'[24] There is no reciprocity, in other words, since 'othering' is based only on consciousness, and the in-itself has none. And thus there is no movement. At best, it is what Jean Hyppolite had said of Platonic alterity – it allows only for an 'immobile dialectic'.[25] Totality is reached once again only through a kind of hypothetical unifying movement – *as if* there were a reciprocity. This then cannot be Hegel's absolute unity or totality, based on the dynamic and generative mobility of *Aufhebung*'s dialectic; it can only be unity in the conditional, a static conditional which never moves to its fulfilment ('detotalized totality', as Sartre calls it).[26] Yet nonetheless Sartre desires it – 'actually we exist on the foundation of this totality and as engaged in it'[27] – and this desire keeps Sartre in an implicit pursuit of a metaphysics on which the foundation of this totality might be 'founded'. It is in this respect that Derrida will label Sartre's concluding intention here a 'metaphysical unity of Being'.[28]

Now it could be argued that, in isolation, an essay on phenomenological ontology does not do justice to the Sartrean corpus, and its thinking on the nature of negation in relation to creativity. For we know that art plays an important role for Sartre in manifesting the nothingness at the heart of being. But if we conclude our own brief discussion of Sartre by turning to his fiction, we find a similar non-Hegelian line. The artistic gesture is a significant part of Sartre's notion of freedom. If being situated in the world is the *sine qua non* of the for-itself, with all its possibilities and correlative nihilations in relation to the 'others' around us, then art presents this situatedness, and its possibilities, in concentrated form. In fact, the imagination of art is part of *the making* of possibility, and so is complicit, deeply if constructively, in nihilation. But with this nihilation comes a responsibility: to act and make responsible decisions in the face of the nothingness that accompanies us. Art is both an acting out (of) the nothing (as imagination) and a responsibility towards the nothing (as doing and praxis). This is why Sartre has such a belief and optimism in the historical and political effectiveness of art in world affairs.[29]

But any such acting is only possible through the self-reflective nature of consciousness. The generative act is part only of the for-itself. It does not exist in any way, even, as we have just seen, in a self-founding way, in the in-itself. Existentially, the creative potency of art does not belong to the nucleus of Being. 'Is' does not 'become'. 'Is' just is. Only when it is conscious of itself in self-reflection can it 'move' out of its fixed, positive state and into the realm of possibility, and freedom. But such a move is clearly not the 'art' we have been arguing for Hegel, an art that arises from the very bowels of Being. And so neither can it be Hegel's art of negation. For Hegel's negation goes beyond simply the nihilation of the fullness and immediacy of one's 'isness'. The art of negation *is* the very arising of 'isness' in the first place, the mediation of its immediacy. And it does not operate out of a certain priority – one can only negate something that already *is there* – but rather it creates the possibility of priority in the simultaneity of its origination: *becoming* as the *Aufhebung* of 'is' and 'is not', being and non-being, immediacy and mediacy. For the early Sartre of *Being and Nothingness*, such an art, the art of becoming, cannot obtain. Such an art would lose all human connection to the world in which we find ourselves.

We have lingered with Sartre in order to show, in a 'negative' way, the turning point he represents in locating Hegel's return. Despite all initial appearances, he is not Hegelian by virtue of his nothingness. He is Hegelian only by virtue of a Hegelianism that would see totality as a metaphysical venture, one that attaches to a positivity, and one that underwrites that positivity in its own presence. But even if that positivity maintains a priority, by refusing any *absolute* reciprocation with its other as negativity, nevertheless, the fact that negativity, as nothingness, encroaches, sidles up to, intervenes in the existent that is being in-itself, shows to us that the Hegelian legacy of negation, however misappropriated, is by mid-century ripening again in a new direction. And that Sartre will implicate imagination in this negation is also not without its significance. To be sure, it is not yet an imagination whose generative qualities are self-generative and whose negative qualities are self-negating. But it is an imagination that acts out the negative, as the opening of possibility and freedom, even if only as a condition and not a source. As Butler puts it, 'Because the world cannot be reclaimed as a constitutive aspect of consciousness, consciousness must set up another relation to the world; it must interpret the world and imaginatively transfigure the world.'[30] This will invite more than just an existentialist ethics onto a scene of general social and cultural negativity, or into the parameters of 'bad faith' as irresponsible action. It will allow creativity once again into the philosophical core of modernity, and incite a rethinking around the relation of creativity to its opposite – dissolution, destruction or erasure. It will only go so far, as we have seen. But perhaps this much was an important contribution to a *Zeitgeist* that would develop beyond itself in the later part of the century, a *Geist* that would, because it must, sublate itself.

JEAN HYPPOLITE – LOGIC AND EXISTENCE

We cannot advance from Sartre's early text into the latter half of the twentieth century without first passing through Jean Hyppolite. Outside Hegelian studies, Hyppolite is known principally as a teacher of later French postmodern luminaries, particularly Foucault, Derrida and Deleuze, all of whom sat under his tutelage at the École Normale Supérieure in Paris, with courses on Hegel. But his *Genesis and Structure of Hegel's Phenomenology of Spirit* (1946) remains a seminal text in revivifying Hegelian interest on the Continent and in France especially (or rescuing it from the dominant anthropocentric readings of Kojève), while his later text *Logic and Existence* (1953) would become highly influential for many of the philosophical issues to come, including difference, alterity, sense, desire, language, the ineffable, structurality and a-logic or non-philosophy.

Negation is generally not included among Hyppolite's legacies, but a brief look at *Logic and Existence* tells us it ought to be. The text is first remarkable in seeing an absolute unity between Hegel's *Phenomenology* and the two *Logic*s (the *Greater* and the *Lesser*). There is less a development and more a logical homogeneity across their space. 'The *Logic* therefore explains the *Phenomenology*', he says.[31] And what it explains is the Logic or Logos of Being in its movement: not a static metaphysics of being, as essence, but an absolute knowledge of being's own self in its fractured existence, a knowledge that being itself must generate. This Hyppolite calls speculative knowledge as opposed to reflective or metaphysical knowledge. 'Being is to itself its own light, its own reflection.'[32]

Hyppolite was able to see that what provides Being with its own power to reflect upon itself is the power that comes from negation. To reveal this, he compares two kinds of negation. The first, empirical negation, does not advance anywhere, because it works from the presupposition and priority of positivity. Here one assumes, as Sartre had done, that being can exist only in a positively ontic sense, and this allows negation to be conceived of only in terms of *exclusion*: it cannot reside in being in any way, and therefore is logically and empirically shunned – being is, while non-being or nothingness is not. The second kind of negation, speculative negation, picks up Spinoza's insight – 'all determination is negation' – and furthers it. It not only admits negation into the very definition of positivity, by apprehending 'the lack or insufficiency in what is present as positive', but it also, at the very heart of this lack, *repeats it*, 'a negation of negation which alone constitutes authentic positivity'. Hyppolite says Hegel's speculative philosophy is a philosophy of negation in this double sense.[33]

The empirical kind of thinking about negation Hyppolite calls a human explanation. For at best it leaves negation to a manner by which we think, to merely *thought* about being or its opposite, non-being. 'Hegel's originality,' Hyppolite writes, 'lies in the rejection of this merely human explanation of negation'.[34] For speculative negation goes to the very core of being itself, a core that, in Hegel, is already wholly united with thought and in thought. Thinking about being is already *to be* in an absolute sense, but with a thinking that is diremptive, and thus a being that is dirempted. For negation, and negation of negation, found thinking and being's very movement and unity, and give them life in that unity. One does not think *about* negation; one's thinking *is* negation. One does not exist *despite* negation; one exists *as* negation. This reorientation to the speculative Hyppolite sees already in the Preface of the *Phenomenology*, in a famous passage (helped to become so by Hyppolite himself) we have already seen ourselves:

> But the life of the spirit is not the life that shrinks from death and keeps itself pure from devastation, but rather the life that endures it and maintains itself in it. It wins its truth only when, in utter dismemberment, it finds itself. It is this power, not as something positive, which closes its eyes to the negative, as when we say of something that it is nothing or false, and then, having done with it, turn away and pass on to something else; on the contrary, spirit is this power only by looking the negative in the face, and tarrying

with it. This tarrying with the negative is the magical power that converts it into being.[35]

Now if spirit tarries, negation does not. Negation, even early on here in the *Phenomenology*, is always active, always in motion. It embraces the other always dialectically, even the other of itself as negation of negation. If Hyppolite was able to help make famous a passage that underlines negation's motive power, he in turn offers a passage that underlines the generative nature of that power: it is not merely motive, but creative; it does not merely set into motion, but creates into being. And thus we note the relevance of the artistic metaphor by which the passage is constructed:

> Platonic alterity allows for an immobile dialectic, a dialectic that still does not have the self for its driving force. Hegelian dialectic, however, deepens alterity into position and opposition into contradiction. This is why dialectic is not merely the symphony of being, being in its measure and in its harmony; dialectic is the creative movement of the symphony, its absolute genesis, the position of being as self. Thus between Platonic dialectic and Hegelian dialectic, there is the same difference as between a symphony heard and the creation of the symphony. The one is being contemplated in its harmony and consonance; the other is the progression of being which posits itself and comprehends itself by positing itself, by identifying with itself in its internal contradiction. This movement expresses the transformation of diversity into opposition, and of opposition into contradiction.[36]

This passage is remarkable on several fronts. First, Hyppolite has just been referencing Plato's *Sophist*, which acknowledges a negative aspect around any determined thing, 'an infinite quantity of non-being'.[37] In trying to resolve the problem this raises for being, Plato, explains Hyppolite, turns the opposite of being (non-being) into something 'other' – non-being is not the negation of being as such but is simply that which points to something other than itself, something different. This puts negation onto a relational plane: what is 'not' are those things that cannot be in any relation to being, those things which stand outside being. And it is the dialectician's task to work out the possibilities and limits of those relations and non-relations. Plato himself uses a musical comparison here to describe this task: 'To possess the art of recognizing the sounds that can or cannot be combined is to be a musician.'[38] But this kind of relational negation, which reduces negation to alterity, is what Hyppolite calls 'immobile'. For alterity remains here with positivity, and positivity does not admit contradiction. Non-being can be something other, but it cannot be itself negating or negated. The dialectic therefore does not move forward or upward; it remains static across the plane of its relations or across the *state* of their alterity. Neither does it penetrate into the very nature of the self, but keeps its alterity forever external. Hegel's dialectic, on the other hand, moves above or through the plane of relations, and *changes* the state. In doing so it moves back towards the very heart of being. Hyppolite says it deepens alterity, first by moving it out of an external state of 'relations' and into an internal 'position' (a move that would, paradoxically, suggest a non-move, a moving to position), and then by moving it from 'opposition' towards 'contradiction' (a move that, again paradoxically, restores its movement).[39] Opposition is not the same as contradiction for Hyppolite. The former positions itself over against another, while the latter moves itself against another through the dynamism of negation. Hegel's dialectic thus radically alters alterity, for it internalises alterity as a position of the self, which in turn becomes its *op-position*, which in turn becomes its contradiction. And with this internal contradiction one is able to move. We recall Hegel's own words in the *Greater Logic*: 'contradiction is the root of all movement and vitality; it is only in so far as something has a contradiction within it that it moves, has an urge and activity.'[40]

From such alteration we can now develop Plato's simile of music, and in doing so come to the second remarkable point of the passage. When Hyppolite says 'dialectic is not merely the symphony of being', or is not 'being in its measure and its harmony', he means that dialectic cannot be reduced to the measure of relations between things, as a symphonic score shows the relations of all instruments, all notes, all sounds to each other in their totality, and as that score is then heard and contemplated by an audience. This would not be dialectical in any motive sense, that is, would not impel 'the creative movement of the symphony'. But neither – and here is the extraordinary 'deepening' – would it be dialectical in the generative sense, for dialectic, as creative movement, is the symphony's 'absolute genesis'. The difference Hyppolite is trying to draw out here is not merely the difference between, on the one hand, the music as a written score heard by an audience and, on the other, the execution of that score as performed by musicians. The difference is just as much between the performing of that score and its very initial creation. Dialectic, as negation, brings into being. It is the very engine of creativity.

A yet third remarkable point is that this absolute genesis is 'the position of being as self'. Being does not simply contemplate itself in relation to its surroundings; being is the progression that 'posits itself and comprehends itself by positing itself', being as begetting itself in its own self-knowledge and 'identifying with itself *in its internal contradiction*'. Being's 'position' then is thus a movement of self-*positing* and self-*opposing*, and so a movement that transforms any positionality (and positivity) of the self into the vital and generating impetus that is contradiction. It is only on this basis that the self can move forward into the absolute unity of itself with its non-self, and into the Absolute itself. But if it is to tarry there in that Absolute, it must do so contradictorily: it must move itself beyond itself. This is the great logic of existence that Hyppolite brings home to us in his astute reading of Hegel, what he elsewhere has emphasised, employing Hegel's own language, as the 'disquiet of the self' – its unrest, its restlessness.[41]

Hyppolite's contribution, then, goes far beyond mere commentary. In advancing us beyond Kojève's humanist and post-historical reading of Hegel, he shows how a proper reading needs also, contra Kojève, to go beyond a fixation with the *Phenomenology* – it needs to include the Logic (as text(s) and concept) at its very centre. But more than this, he shows negation working in the very nucleus of Logic, and therefore in the very nucleus of Being itself. And this negation is not just differential. Or its differential is not predicated upon a difference/identity or relation/non-relation matrix. It is not *predicated* at all, in any sense. Nor is it *inflective*, simply altering the relations within the grammar of ontological existence. It is *generative*, in that it brings to life. It is absolute genesis, creating the symphonic whole. And yet because it is a contradictory impulse, that whole is never whole. It is *Unruhe*: never resting, in both senses of never arresting itself into repose and of never being at peace with itself. It is a dynamic disquiet. And one to have much influence on subsequent thinking, even beyond Hegel.

We have reached that point whereby the Hegel we had set out in Part I can fully come into his own anew. We have repeatedly marked the paradoxical nature of any such move, for the immersion into 'one's own' as its full and proper fulfilment remains at fundamental odds with the generative nascence of the newness we have in play. When we now say Hegel can come into his own, we imply therefore that he can be refuted, rescinded, revoked or *negated*. But he can only be so *properly*, that is, in the nature of his own very self, and his own very thought, by staying with him and remaking him. If we negate Hegel, we negate his own negation. The *negation of Hegel* is always, necessarily, cogenitive.

The cogenitive nature acting upon Hegel, by means of negation's inherent reciprocity with-in Hegel, is also played out in the historical and intellectual narrative we have been suggesting. If in this chapter we have seen a preparation of the ground for a reseeding of Hegel's negation, the seasonal movement is circular and not linear. We circle back on Hegel, as each of the figures – Sartre, Hyppolite and so on – circle back, returning Hegel in their way (in replacement or in renewal). We do not suggest, nor can we suggest, that Hegel comes back to us along the progressive path of a philosophical *eschaton*. Hegel does not stand anywhere as a culmination. The history of Hegel, of his reception and his repetition, is a sowing of Hegel through the self-scattering of his own spores, a kind of self-planting or self-pollination through the diaspora that is his own death. We overstretch the agricultural metaphor here deliberately. Hegel's organicism is precisely the cogenitivity of growth in all directions: the death of Hegel is the birth of Hegel, the birth his death – the *birth of his death*. We return back to Hegel in the fashion of this circularity because Hegel's own thought is the thought of that return. To take him up again, to harvest him again, is to cut him down, and allow him to take root anew, perhaps even elsewhere. Here we recall Foucault's words about trying to escape or go beyond Hegel, in a speech delivered upon the assumption of Hyppolite's chair at the Collège de France: 'We have to determine the extent to which our anti-Hegelianism is possibly one of his tricks directed against us, at the end of which he stands, motionless, waiting for us.'[42] But that end, of course, as Foucault very well knew, would be a beginning.

Those we now turn to, positioned as they are within the latter half of the twentieth century, and the beginning of the twenty-first, are those who have most appropriated this internal paradox in Hegel, who have allowed, even encouraged, its growth and dehiscence within their own thought. They are not an exclusive group; the company that keeps with Hegel's negation, who negate Hegel in the name of Hegel, is ever-expanding. But the following chapters will look at those who perhaps have been most attuned to the generative function of this negation in all its creative capacity. And thus they are those to whom we might look to fashion negation as a new kind of poetics.

CHAPTER 5

The Tolling of Hegel and Negation: Derrida

There is no philosopher and no thinker more cognizant of Hegel's own inner irruption, and of the need to read Hegel according to that irruption, than Jacques Derrida. In all his works about, relating to, touching upon, implying, winking at or glancing sideways to Hegel – and they were many, from his earliest essays through to his last writings on religion, politics, friendship and so on – Derrida was always bringing to the fore, through his own gestures of enactment and re-enactment, the auto-irruptive nature of Hegel's own logic. But he was also deeply aware that such a nature demands of the reader an astute engagement with, and a certain embracing of, that logic. Derrida had written that 'the passage beyond philosophy does not consist in turning the page of philosophy (which usually amounts to philosophizing badly), but in continuing to read philosophers *in a certain way*.'[1] And no more was this the case than with the philosopher Hegel. He had famously acknowledged in an interview of 1971, part of several that became the text *Positions*, of his own position vis-à-vis Hegel: 'We will never be finished with the reading or re-reading of Hegel, and, in a certain way, I do nothing other than attempt to explain myself on this point.'[2] But that 'certain way', which repeats itself, is a going beyond Hegel by lingering with Hegel, a surpassing by means of a loitering, a breaking with by means of enjoining. In that same text, in an interview with Julia Kristeva, he had said: 'I do not believe in decisive ruptures, in an unequivocal "epistemological break," as it is called today. Breaks are always, and fatally, re-inscribed in an old cloth that must continually, interminably be undone.'[3] There is no question of Derrida's own re-inscriptions of Hegel in the very fabric that holds together, and at the same times renders asunder, Hegel's system. He writes of 'relaunching in every sense the reading of the Hegelian *Aufhebung*, eventually beyond what

Hegel, inscribing it, understood himself to say or intended to mean.'[4] Derrida, like his predecessors, is thus forever circling back within the circle that is Hegel and Hegelianisms in order to break out of that circle. But he can only do this by remaining within the circle's orbit, or by spiralling, spiralling inwards or outwards or in both directions simultaneously. This idea of spiralling around – *with-in* – Hegel we will return to below.

To talk *about* Hegel, to philosophise *about* him, in every sense of the word 'about', also involves an obliqueness. When Derrida addresses Hegel, in any of his writings, he does so most often by means of or alongside another voice, or set of voices. He never addresses Hegel directly or exclusively as if he was the only person in the room. Even his most exclusive and sustained treatment of Hegel, his early essay 'The Pit and the Pyramid: Introduction to Hegel's Semiology', begins with the caveat, 'we will proceed chiefly by detours, following texts which more appropriately [but what can Derrida mean by this 'appropriately'?] demonstrate the architectonic necessity of the relations between logic and semiology', texts such as, he tells us, Hyppolite's *Logic and Existence*.[5] That is, he never seems to give Hegel sustained eye contact. This might be because, standing before the one who most, and perhaps first, taught him about difference, he holds a certain shyness and/or reverence (that mixture of responses a person often experiences in encountering, years later, the one to whom and with whom virginity was lost, an image Derrida does not elsewhere shy away from in talking about auto-eroticism, homo/heteroeroticism, the hymen and sexualisation, as in *Glas*, for example). Or it might be because Hegel is never alone himself, because Hegelianism is so amenable to different sets of eyes, different interpretations, different appropriations, that to look at Hegel now, after two full centuries of his thought, is always, and out of necessity, to look equally at other thinkers (Kierkegaard, Marx, Feuerbach, Kojève, Hyppolite, to name only some of the more obvious) as much as other artists (Bataille, Sartre, Genet, etc.). This is to say that, for Derrida, to look at Hegel directly requires indirection, that to allow Hegel to come into his own is to *mediate* him, to read him and re-read him through the mediation of others, and for Derrida this significantly includes the mediation of the artist.[6]

We cannot pretend to do justice here to all of Derrida's work on Hegel. Others have done this collectively elsewhere.[7] Our focus remains on negation, though of course negation itself pervades the Derridean corpus like a brocade of capillaries that goes on endlessly and that one can never extricate without permanent damage. If, however, we limit our attention to the texts and passages on Hegel's negation, we might glimpse how the entire vascular system functions, a system, ironically, not unlike Hegel's: at any entry point into its network one eventually joins the common flow. And the lifeblood of that flow, it bears repeating, is negation.

FROM THE WORK TO THE PLAY OF THE NEGATIVE

Let us begin then with the text from *Writing and Difference* that is one of the most common entry points, 'From Restricted to General Economy: A Hegelianism Without Reserve'.[8] Almost immediately we encounter two of our heralded traits: this essay is as much about another, Georges Bataille, as it is about Hegel; and Hegel is exposed to a fierce critique by means of a deep utilisation. The first will elide with the second. With the first we can say that Bataille, and especially his text 'Hegel, Death and Sacrifice', gives us the framework for the discussion to unfold. But we are never certain whether Derrida wants us principally to read Bataille (as a reader of Hegel) or Hegel (as read through Bataille). Who stands as the framework and who the content within? Who structures whom? Both, naturally, are unfolded in the course of the essay, but in a manner that Derrida wants us to understand as impossible to separate: the framework is itself framed by the content as much as the content is framed by the framework. This elision between form and content and between method and meaning, or their collapse into each other, we have seen as central to Hegel's own logic, and it is why it is so salient here. The collapse emerges in numerous places, variously disguised: 'a complicity without reserve accompanies Hegelian discourse' (p. 253); 'A "principal text" would be the one… which places knowledge "at the height of death"' (p. 254); 'A trembling spreads out which makes the entire old shell crack' (p. 260); '"I write in order to annihilate the play of subordinate operations within myself"' (pp. 266, 273); and so on.

With the second trait, we are told at the outset that Hegel and Hegelianism must be exposed to a certain laughter. At face value we cannot take Hegelianism seriously. But because Derrida never looks straight into the eyes of Hegel in any prolonged manner, he immediately qualifies this claim: any such laughter must issue only from a serious encounter. If laughter is a giving up of control to the convulsions of what exceeds the strictness of propriety and method (in discourse), laughing at Hegel requires stricture. 'To laugh at philosophy (at Hegelianism) calls for an entire "discipline", an entire "method of meditation" [Bataille] that acknowledges the philosopher's byways, understands his techniques, makes use of the ruses, manipulates his cards, lets him deploy his strategy, appropriates his texts.'[9] Is this Bataille laughing at Hegel or Derrida laughing at Hegel, or even at Bataille? However understood – and in some sense it must be all three options[10] – there is a critical impetus in which the abandonment of laughter is possible only because there is a method which brings us to the abandonment (the abandonment of laughter cogenitively). This method is internal: Hegel surpasses, abandons, exceeds himself in all his seriousness, the seriousness of his text. A critical passage on this excess captures Derrida's intention:

> Necessary and impossible, this excess had to fold discourse into strange shapes. And, of course, constrain it to justify itself to Hegel indefinitely. Since more than a century of ruptures, of 'surpassings' with or without 'overturnings,' rarely has a relation to Hegel been so little definable: a complicity without reserve accompanies Hegelian discourse, 'takes it seriously' up to the end, without an objection in philosophical form, while, however, a certain burst of laughter exceeds it and destroys its sense, or signals, in any event, the extreme point of 'experience' which makes Hegelian discourse dislocate *itself*: and this can be done only through close scrutiny and full knowledge of what one is laughing at.[11]

This experience of self-dislocation, this being beside oneself (in laughter), is precisely what it means for Hegelianism, as much as Hegel, to succumb to its own negation.

But now what is negation in this text? Derrida draws out Bataille's reading of the master/slave dialectic (or Bataille's re-interpretation of Kojève's anthropomorphic reading) to seize upon Bataille's idiosyncratic understanding of sovereignty. In the master/slave passage, the master can never fulfil his own self-consciousness, since it demands recognition from the slave in order to be fully actualised. The master is dependent. The slave, despite his servile role, or more precisely because of it, comes to recognise his labour as his own achievement, and though it is on behalf of the master, and is in constant need of the master, that very recognition frees the slave to his own self-consciousness, and eventual freedom from and overcoming of the master. Whereas the lord's self-consciousness first requires recognition *from* the other, the slave's self-consciousness enacts recognition *as* the other, and this enactment is the liberating moment that allows the slave his own mastery.

Now for Derrida, following Bataille, the entire dialectic of life-and-death struggle between master and slave is geared towards mastery. Even if the slave is willing to put his life on the line, he does so in the hope of commuting his servility and his death, and thus – in the history of this passage's interpretation since Marx – for the purposes of what Derrida calls the 'servile condition of mastery'.[12] The negation traditionally involved here, whereby the one is struggling to negate the other, but where only the one who can recognise the internal self-negation can be truly free to negate the negation, is still a negation caught up within a certain pre-established horizon. This is Derrida's fundamental criticism of Hegel's *Aufhebung* both here and in virtually all his other texts where Hegel figures: the sublating that operates within the Hegelian system is always and already *teleological*. It can never surpass its own trajectory, since its vectors are always factored within the pre-ordainment of its system. It is always factoring negation towards presence and meaning.

In Bataille, Derrida finds a possible way out of this systemic problem, a way to exceed Hegel at his own game. It is with the notion of sovereignty that does not *master* itself, even in its own impoverishments, but escapes the polarity of master/slave, mastery/servility altogether. Sovereignty here does not mean reigning over, but a retreat to an absolute position (though Derrida tries – not wholly successfully – to steer clear of the term 'absolute' in this connection) that has no relation whatsoever with the symmetry and dissymmetry of master/slave. This is a sovereignty that is not a state or a concept, or even a concept of a state of rule or mastery, but rather a sovereignty that '*does not govern itself*'.[13] It is an operation that constantly slips away from the government of meaning and from meaning itself, and therefore becomes a 'sovereign silence'. It is, we might say, an early articulation of what is to become deconstruction, the sovereign silence 'foreign to difference as the source of signification' and 'to a *system of meaning* permitting or promising an absolute formal mastery'.[14] Such a sovereignty would be a different kind of negativity – not one that allows a movement towards mastery, but one that exhausts itself wholly in its own movement, *without reserve*.

The ruling context in which both Bataille and Derrida read and construe Hegel's dialectic here, and the sovereignty that might surpass it, is labour. The slave works and produces, the master sits back and consumes. The negativity between them is one of expenditure: the slave negatively expends his work (it is not for himself but another); the master negatively consumes that which is expended (it is not his product). Or, the slave negatively expends himself for the other, the master negatively expends himself in his desire for recognition from the other. It is in this sense that negativity is interpreted 'as labor' that always maintains a reserve: there is something held back in the expenditure that allows it to keep expending, whether the reserve to keep producing more, the reserve to keep desiring recognition, or the reserve to overcome. And it is that reserve that supplies the means to mastery, because it keeps present the resources for meaning, for discourse, for aims, for teleology, in the face of an absence that is the death to which the dialectical struggle exposes the combatants. It allows death a meaning, revolution a purpose. As Derrida exploits the image of labour economy: 'The notion of *Aufhebung*... is laughable in that it signifies the *busying* of a discourse losing its breath as it reappropriates all negativity for itself, as it works the "putting at stake" into an *investment*, as it *amortizes* [with etymological significance] absolute expenditure; and as it gives meaning to death, thereby simultaneously blinding itself to the baselessness of the nonmeaning from which the basis of meaning is drawn, and in which this basis of meaning is exhausted.'[15]

The Hegelianism *without reserve* of the title, an *Aufhebung* that would not keep in reserve its negativity for positive ends, but would exhaust itself thoroughly and comprehensively, without any remainder, is what Hegel cannot see, according to Derrida's charge. 'The blind spot of Hegelianism, *around* which can be organized the representation of meaning, is the *point* at which destruction, suppression, death and sacrifice constitute so irreversible an expenditure, so radical a negativity – here we would have to say an expenditure and a negativity *without reserve* – that they can no longer be determined as negativity in a process or a system.'[16] Sovereignty, in Bataille's terms, would be this *'point of nonreserve'*, which is 'neither positive nor negative'. Derrida says that Hegel not only has a blind spot in regard to this point, but that he wilfully 'blinded himself' to it. Yet this means that he in some sense became aware of it, glimpsed it, maybe even understood it. He laid it 'bare under the rubric of negativity'.[17] He even called it something: 'abstract negativity'. But in that abstraction he diminished it, exiled it, defanged its transgression. This is why Derrida says, quoting Bataille, 'he did not know to what extent he was right.' He was 'wrong for having been right, for having triumphed over the negative', by binding it forever to the underside of positivity, as 'the reassuring *other* surface of the positive'.[18]

The Derridean need to overcome classical binaries is here clearly seen. Hegel, like Kant before him, sees the negative for what it really is, but pulls back, and keeps it serviceable within a binary opposition. For Derrida, reading Bataille, this restricts it to an economy of meaning and mastery. It disallows the 'play', the 'sliding', the 'displacement' that the 'nonreserve' imparts. It disallows it 'to accede to the sacred' (which is an entire other matter Bataille's reading invokes, and which we must reserve for elsewhere). But there are several factors that restrict Derrida's own reading. The first is that by construing negativity and the *Aufhebung* only in terms of labour and expenditure, he limits the potency that is negation's ownmost nature. Negativity, he says, becomes a *resource*; it is not a *source*. It is there to draw from; it is not there to initiate. So it operates from an economy, and one necessarily restricted. Second, that economy must always be political – one never labours outside the construct of power relations – and therefore is bound to an oppositional framework – a binary – of have/have-not, of owner/worker, of free/enslaved, of positive/negative and so on. That framework may invite a deconstruction, but it tends to suppress or rein in any sense of becoming. Third, and as a result, that framework trades more in the nominal and static sense of the negative ('negativity') rather than its verbal and active sense ('negation', a term seldom if ever used in this essay). And lastly, whenever the negative is understood in its more active sense, say in the way both Kojève and Bataille conceive it – negativity as 'Action' – it is towards a teleology, a specified consummation in something positive and present.[19]

From our foregoing chapters, we know that this is not the only way to read the negative, and not even its most proper way, if by this we mean true to its ownmost nature. What we have called the art of negation shares many features with what Derrida, appropriating Bataille, has called sovereignty without reserve. It already operates outside the economy of binary oppositions, and certainly outside the economy of mastery (of meanings, of the other, etc.).[20] More, it operates with the kind of play, the kind of abandonment of laughter, that Derrida claims is part of the excess of its operation. But it is not just an operation, a work, or even an ergonomics; it is an origination.

Now Derrida himself glimpses this in Bataille. In a single central paragraph, after juxtaposing 'discursive signification' with the irruptive opening of sovereignty that goes beyond the limits of discourse and absolute knowledge, he acknowledges that Bataille 'sometimes opposes poetic, ecstatic sacred speech to "significative discourse"', and that, though it does not transcend discourse altogether, the poetic nevertheless 'is that *in every discourse* which can open itself up to the absolute loss of its sense, to the (non-)base of the sacred, of nonmeaning, of the un-knowledge or of play'.[21] But too often poetry is subordinated to the cause of meaning, or is inserted back into metaphorical domains of mastery and servility. It therefore yields to the positive. It is only 'within the interval between *subordination, insertion*, and *sovereignty* that one should examine the relations between literature and revolution'.[22] But that is all Derrida says of it. He does not develop this poetic further here. Even when he asks if Bataille himself had ever found a sovereign form of writing, he only speaks, perhaps with Blanchot in mind, of a unique and sometimes scandalous 'space of writing' that Bataille forever pursued. But he, Derrida, never explicitly mentions the poet or artist again in this essay. Like Hegel, he has glimpsed the art of negation briefly, but retreated. Only in subsequent texts will that glimpse be opened up into a more generative understanding.

THE VESTIBULE OF NEGATION

What was being worked out in the 1967 essay under the directions and insurrections of Hegel and Bataille comes to form many of Derrida's lifelong pursuits under different terminology and with different sets of debaters. Here as much as anywhere, the entry points find their way into the bloodstream of his overall work. Let us take these ideas of the 'poetic insertion' and the 'space of writing' with which we have just concluded to show how Hegel returns in a more disruptively creative manner, which will lead to what might be called Derrida's *Finnegans Wake*, the text most inaccessible and most unreadable to the average reader of Derrida, *Glas*.

Written in 1974, *Glas* was one of several full-length books Derrida experimented with during the 1970s, before he returned to the essay format in which he remained most prolific.[23] Like these other monographs, it brought together philosophy and art in the most unconventional of manners, and indeed layouts. But it had precedence, or at least a dress rehearsal, with an earlier essay entitled 'Tympan', which introduced his 1972 volume of collected essays *Marges de la philosophie* (*Margins of Philosophy*).

The volume's title was indicative enough. It centralised his fascination with limits, borders, framing and empty spaces which had already come to light in the previous volume *Writing and Difference*, and in the Hegel/Bataille essay itself. What is first remarkable about the opening essay 'Tympan' is its non-standard typographical layout: two columns of text are set side-by-side, the left two-thirds the width of the right (at least in the English layout), each with its own seemingly independent narrative. The larger is Derrida's text, a kind of introduction to the volume to follow, which of course includes 'The Pit and the Pyramid: Introduction to Hegel's Semiology'. The smaller is an excerpt from a text (*Biffures*) by Michel Leiris. What is Leiris doing introducing a text about a volume dedicated to philosophy? Granted, Leiris was a marginal philosophical thinker, but the passage chosen has no direct and apparent relevance to philosophy itself. It speaks, and in a broken manner, of plants and shells and earwigs and vocal cords and old gramophones. Why this inserted column of Leiris? Why the splitting of the text?

The title page of the essay presents three epigrams from Hegel. The first is from the *Science of Logic*, and speaks about the limit that stands between any thesis and antithesis, but is itself sublated to go beyond itself, and beyond any limit. The second and third, from the earlier *The Difference Between Fichte and Schelling's Systems of Philosophy*, speak of beginning or attaining philosophy either by the necessity of philosophy to cast itself into an abyss *à corps perdu* (lit. with a losing of the body, or abandonment – like laughter) or by it offering a kind of vestibule (*eine Art von Vorhof*) for itself, with which it can enter itself. Without explicating these images too far (even if they deserve expansion), we can see how Hegel once again supplies us with the question of surpassing the limit by necessarily going beyond it, and of creating the ground or the entry of its own operation. And so it is that Derrida's column in 'Tympan' opens with reference to Hegel's *Aufhebung*.

A full exegesis of this short opening essay would take us too far into all the intricacies at play between Derrida, Hegel, Leiris and the nature and limits of philosophy. But let us draw out at least some the most prominent features for an understanding of philosophy's limit (and Hegel's limit) in relation to creative negation. The 'tympan' of the title originally refers to a thin diaphragm or barrier that separates one space from another, and upon which reverberations might be sounded (as with a drum or tympani). Such a diaphragm features within the ear canal, as part of the human auditory apparatus, the ear drum. Not only does it allow us to hear things, but it acts as the separating limit between the exterior world and the interiority of our head. What happens in the vestibule of the ear translates into the thoughts of the mind. This limit is where modern philosophy begins. '[E]verything comes down to the ear you are able to hear me with', Derrida says later when he returns to the limit of the ear in an essay on autobiography.[24] Hegel's *Aufhebung* draws attention to this limit, because it is precisely at the limit (the *Logic* tells us) where the *Aufhebung* operates. The other (the outside to the inside, or the inside to the outside)[25] is there confronted, and there surpassed. But thinking the other to itself, *philosophy's* other, threatens to break through the limit of philosophy and carry one beyond philosophy. If philosophy truly begins with itself, but itself as its *other*, then where does it take itself? As philosophy's own vestibule, where does the *Aufhebung* (and let us remember this is a quintessentially *philosophical* word and concept) lead? It leads to its margins, and to an abyss.

Leiris' text, a more artistic text (is it prose? is it poetry?), presents one possible *other* to the philosophical text. It too speaks of ears and ear canals, but in a very different fashion, and with (presumably) a very different purpose. (Rent from its original context, we cannot deduce that purpose, which is the point.) It also presents a margin or a gap no longer on the periphery, but as an obtrusive *interior* separation between two kinds of writing, one supposedly philosophical, one supposedly not. This space is an insertion, an intrusion. It disrupts the normal limits of thinking, and reading. What does one do with this margin? For the columns erected across either of its sides are perfectly parallel but never tangent; their narratives never intersect directly, or flow from bottom to top as in the typography of a newspaper or magazine. They remain distinct, wholly other to each other.

But of course they are not wholly unrelated. They touch each other – they touch the same images and metaphors and even occasionally ideas. But they touch only at a blank limit and separation. Like the ear canal, their propinquity is separated by a transparent diaphragm, in this case an empty space. Like the tympan, one reverberates upon that space to transfer over to the other. The hammering at the margin of the one side allows the other side to be heard, to be taken up. But it only does this by way of the abyssal, empty nature of its blankness. A possession through separation and nothingness. What is a barrier is also a passageway, but one we cannot 'see' with philosophical eyes, one we cannot philosophically appropriate, because it exceeds philosophy proper. So 'Tympan' becomes an improper philosophical text. And even more so the *Glas* which models itself upon it. 'To philosophize *à corps perdu*. How does Hegel understand that?' Where does the body of a text and of a philosophy go when it butts up against its own internal margin? So Derrida asks of 'Tympan', at the vestibule of his volume, 'Can this text become the margin of a margin? Where has the body of the text gone when the margin is no longer a secondary virginity but an inexhaustible reserve, the stereographic activity of an entirely other ear?'[26]

There is one other image here we can now return to: the spiral. The spiral is part of the geometry of the inner ear (the cochlea); Leiris' text draws our attention to this. In fact, the passage opens with numerous instances of the spiralling shape in nature – botany, the shell of a snail, curls of hair and so on. Above we talked about the circularity of the Hegelian return, of the Hegelian text and of Hegelian negation. What is an end is also a beginning, what enters also exits, what is logical is also *tauto*logical and so on. The motion of the circle is two-dimensional, and what goes around comes around. But the motion of the spiral or helix coil adds a dimension. In the case of the helix (a fixed diameter), the circle advances upwards or downwards around the abyss, leaving us uncomfortably on the circumference. In the case of the spiral (diminishing diameter), the circle draws us inwards to or outwards from some infinite and impossibly unknown centre. Derrida reads Hegel's *Aufhebung* as a spiral that, despite its revolutions around the negative, draws us outward, away from the centre. Derrida reads deconstruction as a spiral that leads us back into it, the operation of the *Aufhebung* turned in on itself (not just reversed, but turned outside-in). This has a dizzying effect for the philosopher, to be sure. If Bataille's hunch was right, the poet, the artist, might be better adjusted to such a movement.

From this we can now move to *Glas*, but by one further detour or passageway: ironically the one essay most directly devoted to Hegel in *Margins of Philosophy*, 'The Pit and the Pyramid'. Most of this examination of Hegel's understanding of semiology concerns the nature of signification and writing, and it would require an elaborate and prolonged analysis to translate this examination into terms directly pertinent to questions of negation, far beyond what we can afford here. But one passage stands out as a direct bridge from 'Tympan' to *Glas*: the discussion of Hegel's paradoxical notion of 'physical ideality', as manifested in the senses of sight and sound. Sight, Hegel tells us, is not physical in the same way touch and taste are. Like the sign, sight cannot be eaten. But if sight has an ideal aspect to it, hearing does even more so. 'It "relifts" (*relève*) [sublates, *aufhebt*] sight'. Sight retains the exterior presence of the things that are seen – 'for example, plastic works of art' – and therefore such things 'resist the *Aufhebung*, and in and of themselves cannot be *relève* by temporal interiority. They hold back the work of dialectics.'[27] But music and speech are different. Things heard are not retained in their exteriority in the same way. They are more thoroughly inwardised. This makes hearing, according to Hegel, more 'theoretical', more 'ideal'. And Derrida quotes a long passage from the *Aesthetics* in which Hegel speaks of sound's 'double negation'. We will quote only a portion of this passage, as itself a tympan between 'Tympan' and *Glas*, and let it reverberate:

> The ear, on the contrary, without itself turning to a practical *(praktisch)* relation to objects, listens to the result of the inner vibration *(innernen Erzitterns)* of the body through which what comes before us is no longer the peaceful and material shape, but the first and more ideal breath of the soul *(Seelenhaftigkeit)*. Further, since negativity into which the vibrating material *(schwingende Material)* enters here is on the one side the *relève* (*Aufhebung*) of the spatial situation, a *relève relevé* again by the reaction of the body, therefore the expression of this double negation, i.e. sound *(Ton)*, is an externality which in its coming-to-be is annihilated again by its very existence, and it vanishes of itself. Owing to this double negation of externality, implicit in the principle of sound, inner subjectivity corresponds to it because the resounding *(Klingen)* which in and by itself is something more ideal than independently really subsistent corporeality, gives us this more ideal existence also, and therefore becomes a mode of expression adequate to inner life (*Aesthetics*, II, p. 890).[28]

We'll only add that Derrida notes the prevalence of the idea of vibration or trembling, both here (*Erzitterns*, as above) and elsewhere: 'In the *Philosophy of Nature* it is at the center of the physics of sound (*Klang*); and there as always, it marks the passage, through the operation of negativity, of space and time, of the material in the ideal passing through "abstract materiality"'.[29] The 'tympan', as the transparent diaphragm of the limit, or the margin, is as much this operation of negation as the sound (*Klang*) it allows itself. And *Glas* becomes this sound (as *Klang*).

THE SOUNDING OF NEGATION: *GLAS*

Even more than 'Tympan' and 'The Pit and the Pyramid', the intricacies of *Glas* are far beyond our present scope. They are beyond most scopes, except for a multi-volume exegesis of Joycean proportions.[30] It would be folly to attempt or to expect an exhaustive analysis of the content here, or even a summative analysis. Others have done this, if not yet comprehensively.[31] Instead, we will concentrate on the form of the text, and by doing so, show once again how form and content are interrelated, mutually inclusive, reciprocal or, perhaps most significant of all, products and offspring of each other.

What was a typographical rehearsal in 'Tympan' becomes a full-blown performance in *Glas*. Typography, we might say, is wholly rewritten. It is no longer simply two columns, one (the philosopher Derrida) two-thirds the size of the other (the writer Leiris). If there are two dominant columns, they have equal breadth, and the margin separating them now occupies the centre of each wide page.[32] But each of those columns are themselves cut into, or 'striated', by other text, either as excurses or break-outs. The overall visual effect can be disorientating to the eye, accustomed as it is to reading only in certain prescribed directions and along certain prescribed pathways. But, as Derrida repeatedly tells us in the course of his discussions, what disturbs the eye also disturbs the mind. The sight/site of the written word is closely affiliated to the logic and its meaning. This was the entire point of the analysis in 'The Pit and the Pyramid'. We are logically and philosophically thrown, then, when we encounter the broken, seemingly fragmented, continually striated typography that is *Glas*.

We are even more struck by the ubiquity of the margins, or of marginalising. For now the margins are both central and internal. They become caesurae and lacunae within our thought processes, since we have no direction, no signposting, for how to negotiate around, between, over and across them. The questions of starting and stopping, of linear progress, and of narrative development are all discombobulated. We do not know how to access the text as texts, or to appropriate the text(s), other than by fits and starts. We can only immerse ourselves somehow, but any immersion means an immersion into the blank spaces of the margins as much as into the cubist-like arrangement of the written words. Margins here become both the liquefaction and the liquidation of the written text.

Moreover, Derrida has confused his own authorship. Whereas 'Tympan' very clearly separates one voice from another (Leiris was only ever speaking into the ear of Derrida), here both – we shall say *all* – columns are reverberating and resonating with a cacophony of voices and authors, as if each were a competing tower of Babel. Of course, Derrida's voice must be heard over all of them, but the confusion of tones and intonations makes them difficult to distinguish, and the quotation marks often indicate more of a vibration of sound, the plucking of someone else's strings, than a distinct or direct speech. The 'defect of direct form', as Derrida quotes Kierkegaard speaking of Hegel.[33]

And finally there is the question of the columnisation itself: the erecting of blocks to form little towers, obelisks, edifices or monuments upon each page. But their broken nature suggests a fragile solidity, or a ruination of some past glory, as in the depiction of classical ruins in Renaissance art. ('Remains' figure in the first sentence of both columns on the opening page.) Or they suggest a phallicisation of the text gone somehow flaccid, or rather of its spermatozoa spilt before penetration could occur. And of course, these images are exactly what Derrida wants to suggest. And more.

If we try to bring 'order' to this arrangement, we can without too much difficulty determine the left column (with its cuts, insertions and caesurae) as Derrida addressing and quoting Hegel, and the right column (with its cuts, insertions and caesurae) as Derrida addressing and quoting Jean Genet. This 'duality', this 'tympanic modality',[34] immediately sets up an interrogative posture we saw in 'Tympan': why are these two figures set side-by-side, and how do we read across them? The philosopher stands on the one side, the artist on the other, and between them lies an invisible or transparent membrane in the form of a centralised margin. Is this central gap a boundary and limit or a threshold and passage? How do we confront its presence, negotiate its space? But the tympan has already instructed us: we occupy one side while the other hammers on the skin of the partition, to send a reverberating signal to the other. But then a further question intrudes upon us: on which side is the sound's source, and on which side its physiological conversion? Which is the exterior, and which the interior? Where are nature and the material, where the mind and the ideal?

We know the Hegelian answer. If it is really a question of art on one side, philosophy on the other, we know these two are more than affiliated: they are identical in content, even if not in form. Philosophy forms itself internally, as the logical working-out of the concept, while art forms itself externally, as the materialisation of that concept. The artist creates the music, the philosopher gives it meaning. Dialectically, the art on the right-hand side is surpassed by philosophy on the left-hand side.

But of course in Hegel's scheme, between stands religion. To get from one to the other we need, formally, to go through religion. But can religion here be the blank gap in the middle, the sovereign sacred that Bataille had spoken of? Yet religion, in Hegelian terms, is also mediation; at best we can speak of religion as a form of art, or art as a form of religion. Since Derrida's left-hand Hegelian text opens and closes with religion in the form of art, and speaks continually of religion throughout, does this mediation itself get subsumed by philosophy, in good Hegelian fashion? But we are still left with the empty space in the centre. What are we to do with its blankness?

Let us return to the title, the word 'Glas'. In French, this means 'bell' or 'knell', and, relatedly, 'sound'. In German, the word can also mean 'bell', but more commonly means 'glass'. A bell functions in the auditory realm, while glass in both the material and the visual. If the operation of the bell takes us back to the 'tympan', this is because the clanging of the bell, like the hammer on the diaphragm of the inner ear, involves a transmission. How does art penetrate into the depths of philosophical thinking? It transmits its sounds through the verberations generated on a vestibule membrane, which allows the conversion from the physical to the ideal. We have already seen that sound has this privileged halfway nature: it is more ideal than even sight, since its substantiality is only preserved in the mind of the hearer.[35] But where does this conversion take place? It takes place on the threshold where the exterior is now interior but the interior is still exterior, and where the empty canal is partitioned by a transparent boundary. The centre margin thus becomes the liminal space where interior and exterior collide, and disappear into each other.

In the more common meaning of the German, the material of glass is also a halfway phenomenon, but now between nature and culture. Glass is generally not found in the world naturally; it is made up of natural substances, principally silica in sand, but forged through extreme heat – or blown (Derrida will employ the French *souffle* frequently throughout his writing, including *Glas*, to conjure the life-giving breath that animates or inspires). The result is a conversion from multiplicity (grains of sand) to singularity, and from opacity to translucence and transparency. Glass also functions halfway between barrier and opening: the window seals, but also allows light and view. Moreover, under the right conditions, the window reflects, showing simultaneously the outside and the inside. So too, art is constantly looking in at philosophy, while philosophy is looking out at art. Both are reflecting each other and themselves, so that the reflection becomes a superimposition of one upon the other, the reflection and the reflected being confused (con-fused) in both directions. Here there is no tain of the mirror. One sees right through, but right through one's own reflection.[36] We return once again to a cogenitive phenomenon: one side is never the preposition of the other. Both are generated on the same surface, a surface that, in itself, is without content.

So the title is indicating to us the nature of a relation. How do philosophy and art relate to one another exactly? On what ground, on what surface, in which space is their connection? Is it conjugal, two separate entities coming together in unity? Or is it filial, one having arisen from the other? If it is conjugal, how do the two sides conjoin and conjugate each other, and what does their conjugation give birth to? Or if it is filial, who is the parent, who the offspring?

The formal structure of the text, then, in questioning the relationship of two sides, carries us through to the content that each side, in their own manner, will treat: the question of family relation or family resemblance, beginning with that between philosophy and art. The opening sentences begin respectively, 'what, after all, of the remain(s), today, for us, here, now, of a Hegel?', and '*what remained of a Rembrandt torn into small very regular squares*'.³⁷ So then the resemblance or relation also stands between form and content: how is the one seen in the other?

To answer this, let us now shift towards the content. The question of family resemblance and family relation is worked out across several recurring themes in the text: flowering, the phallus, begetting, semantics, etymology, religions and death. There are more, of course, but these are among the most prominent.³⁸ Derrida tells us early on in the Hegelian column that he is going to come at Hegel from a very particular angle, 'the law of the family'. In many ways, this approach concerns beginnings, the beginning of Hegel in a return to Hegel, 'the circle of the Hegelian beginning', which involves 'Hegel's family', 'the family of Hegel' and 'the concept [of] family according to Hegel'.³⁹ Cogenitively, one wonders where to begin with the family of Hegel. Derrida begins, significantly, with religion: the religion of flowers as discussed in the *Phenomenology*, and the phallic column of India as discussed in the *Aesthetics*. Two texts, at the beginning and end of the 'system', each seminal in their own way. Seminality is at the beginning of *Glas*, then, both in the case of flowering (germs, seeds) and the phallus (semen, sperm, insemination). But seminality quickly finds its way to the core of Hegelian thought: 'The figure of the seed… is immediately determined: (1) as the best representation of the spirit's relation to itself, (2) as the circular path of a return to self.'⁴⁰ For we are really talking about self-begetting, and how we conceive *ourselves*.

So often we have seen how Hegel returns us to this self-begetting. Being does not simply exist in positivity and plenitude. Being generates itself into being, and this through negation, negation of its own being. Here, Derrida captures texts in which the *human* being is distinguished from the rest of nature, especially the plant and the animal. With the plant, for example, the seed carries its own circularity, for in the life cycle of the flower the seed is both the beginning and the end, both what generates the flower into existence and what the flower produces upon its demise. But this circularity promotes only a self-contained growth: the plant can only generate itself as that plant, as that species of plant. It cannot generate anything outside itself. Its growth, its seminality, its self-mobility, is therefore finite, restricted to the non-freedom of its own nature. The animal too has no more freedom in this regard. But the human being is different, since the human being has consciousness, and in this consciousness is its own self *as freedom*, the ability to create something outside itself, not wholly self-contained. Consciousness is constituted by the awareness of oneself not merely as a thinking being, but as a being who is not fully self-contained, not yet fully formed, and whose freedom allows the potential for development and formation outside its own immediacy. (Here Sartre's reading of Hegel remains in proper allegiance). Yet to think beyond oneself, in freedom, requires a breaking with that immediacy, a negation that divides or renders the self asunder, in order for something other to arise. This negation is a movement *away* from its own natural growth or self-reflexive mobility, as in the plant and the animal, and towards a spirit that has its own other within itself. This 'within itself', however, can only ever be negatively related, since it must exceed itself, surpass itself. 'No longer having, by the fact of this division, its natural movement in itself, it constitutes itself by its *Bildung*, its culture, its discipline, its symbolic formation. Paradoxically, it is, more than the plant or animal, its own proper product, its own son, the son of its works.'[41]

The family is part of this constituting movement within/without. The human family is not just a filiation, but a relation of spirit. (Derrida calls this a 'speculative family'.)[42] The parent relates to the child not simply as offspring, but as spirit begetting itself towards the infinite. One does not simply give birth to the child; one cultivates relations within the familial structure that then, in the Hegelian scheme within the *Philosophy of Right*, lead to the cultivation of the citizen and the State. Family is part of *Bildung*, that German concept of development through education, maturation, culturalisation and spiritual formation. In this sense, animals do not have families; only self-conscious humans, in their freedom, can work towards this kind of spiritual development. This is the development of the *Phenomenology*, as it reaches its heights through the mediations of art, religion and philosophy. The family is an exemplary set of relations that cultivates the spirit within itself and beyond itself. It is the coming together of the finite and the infinite.

Derrida relates this familial enjoining, as Hegel does, to the religious or theological domain, and particularly now (having set up two 'earlier' religious traditions of the flower and the phallus) to the familial structure within the Christian concept of the Trinity. The Father relates to the Son not as offspring; the Father is in the incarnate Son as the incarnate Son is in the Father, the finite and the infinite coming together. But in order for this contradiction to hold, the Father must abandon himself to His other, the Son, as the Son must abandon himself to His other, the Father. In this abandonment, the other is retained as one's own, Father and Son uniting in One. Such retention is only possible within a sublation that is negation, and this sublation is the Spirit. 'The spirit is the element of the *Aufhebung* in which the seed returns to the father.'[43] The self-abandoning is also the self-begetting, which is the spirit of the movement across the family relations.

Derrida is much more painstaking in his analysis of all these relational movements and transactions. But we can see emerging, even in this cursory pass-over, the ultimate relations Derrida is hoping to lay before us: that between art, religion and philosophy, that between form and content, and that between the centre and the margin. If philosophy sits on one side and art on the other, typographically, religion allows us to imagine them as a family resemblance or, further, within a family relation. (The Genet column is also, in its way, weaving in and out of the religious thematics being worked out across the margin.) So, in the midst of the left column, amid a discussion on the Christian family, Derrida inserts: 'One column in the other.'[44] Unlike many commentators, who have tried to make sense of the double nature of the columnisation (a binary opposition, a psychodynamic splitting, two phalluses, even, as Spivak conceived of them, two legs in the middle of which works the phallus to inseminate),[45] Derrida wants us to see them as collapsed into each other, or superimposed upon each other. And religion operates to mediate this collapse. As a Hegelian text, it is never about two things; a third always is operating, even if not immediately detectable. The 'spirit' of the text is in the *Aufhebung*.

Or another way to say this, more ostensibly, is that the spirit of the text is in the partition. The gap in the middle, the central margin, that which partitions philosophy and art, is the religious spirit that mediates the two, or the spirit that is in religion as *Aufhebung*. In religious terms the 'Jewish principle' of a radical separation between human and divine, between thought and actuality, between rationality and the sensible, is overcome, Hegel tells us, in the Christian, or Johannine, understanding of *Logos*, whereby the divine Word becomes flesh. But this *Logos* is already split, says Derrida, following Hegel. The *Logos* is a principle, we are told, not an emanation. God does not emanate as the Son, as 'the continuous production flowing from the source naturally', since under any such emanation He would remain thoroughly divine.[46] Rather, God is the *Logos*, which, as principle, maintains the division or partition between the divine realm and this material world.[47] But as a principle it also serves as the conceptual unification of these two sides in an absolute oneness: in the beginning was the *Logos*, and the *Logos* was God. And yet Derrida says this 'apparent contradiction is thinkable only through the familial determination of the concept of emanation.' Emanation produces, begets, a familial tie: issuance in a relationship of likeness and/or sameness with its source. But how can one uphold the necessary division in this relationship (divine/human), without collapsing its filiation altogether? This ancient theological dilemma (with its debates around *homoousia* and *homoiousia*) is here resolved by means of internal partition: what emanates from one to the other, from the living (father) to the living (son), is not the absolute living of the divine being (what Aquinas, following Aristotle, would call the Pure Act) but instead the partition or division itself. *The emanation of the partition*. The split is always and already part of both sides. This means that emanation, as spirit, the spirit of the text, 'lets itself be worked (over) by discontinuity, division, negativity', so that now '[l]ife and division go together', and 'life is at once the part and the partitioning, the morsel and the whole, its own proper difference, its own proper self-opposition.'[48]

The centre partition is simultaneously life, division and negativity. Which is to say it is also death. If the two sides come together by means of some living connection – the *spirit* – that connection also carries the spirit of death. Derrida writes: 'Between the two lives, as their hyphen {*trait d'union*}, their contract or their contraction – death.'[49] Thus later Derrida will speak of the *Aufhebung* in terms both of copulation and of death.[50] And this is why that death, in its partitioning, also crosses the partition of its own text, and plays itself out on either side, uniting philosophy and art. In the Genet column, directly opposite the discussion of an *Aufhebung* that cancels out any plus or minus in the calculus of calculability, Derrida writes that Genet's essay 'Le Funambule' ('The Tightrope Walker'):

> is divided in two: paragraphs in italics, paragraphs in roman, narrative {*récit*} and apostrophe. I-it/he, I-you. The wire of the text disappears, reappears, stretches itself to the point of vibration, becomes invisible through too much rigor or too many detours, loads itself with all the names, bears Death and the dead man. Already – the vigil {*veille*} – the place of the dead man: 'Death – the Death of which I speak to you – is not the death that will follow your fall, but the one which precedes your appearance on the wire. It is before mounting the wire that you die… But see to it {*veille*} that you die before appearing, and that it is a dead man who dances on the wire.[51]

The theme of death as falling itself falls, like the pealing of a bell, throughout the text, just as the theme of spirit as rising resounds from page to page and from column to column. Death and spirit are no more separated as they are in the image of the tightrope, which is strung up high above a perilous drop and across a perilous expanse, oscillating up and down as the funambulist proceeds across. Their unity is already foregone, just as the death of the funambulist is already foregone. But it is precisely in that death that the dance of the feat is possible. This dance is a 'religious act', Derrida later suggests, an act simultaneously of and against gravity, and he quotes Genet's words: 'The world is attentive to all my movements, if it wants me to trip up. I shall pay dearly for a mistake, but if there is a mistake and I catch it in time, it seems to me that there will be joy in the Father's dwelling. Or else I fall'.[52]

Derrida, through Genet and his image of the tightrope walker, is ultimately portraying for us a family resemblance between life and death, one he has already seen in Bataille, and one he now factors into the threefold Hegelian schema of art, religion and philosophy. But any resemblance is predicated upon a deep relation: it is not merely that one side can see itself in the face of the other; it is that both sides must stand in a fundamental and ongoing relation to one another. A speculative relation. That relation, we can now say, is one of negation.

In the final pages of *Glas*, Derrida returns to Hegel's treatment of religion in the *Phenomenology*, particularly where natural religion moves to religion as art. Natural religion begins with God as light, when the divine takes no definite form and is infinitely dispersed across all of Nature. This stage gives way to the benign stage of flower religion, which in turn gives way to the warring stage of animal religion, both plants and animals being the object of worship before any human investiture is made. Derrida focusses his discussion at the point where the human enters religion in the form of the artisan or artificer. At this stage, the spirit of religion moves from an external object to an internal creativity. At first, that creativity expresses itself in straight or angular lines, such as in the pyramid or obelisk, with their plane surfaces and equal parts. These are lifeless expressions in which the spirit cannot yet recognise itself. But then the curve or roundness is introduced, and the spirit begins to come into its own. It starts with columns – straight lines with rounded bodies. But these erect formations give way to more organic forms, in which curvature features more comprehensively, that is, by giving shape through internal growth.[53] Plants and animals are the first to be depicted, and later the phallus, the organic column with generative potency.

But interwoven within this discussion of religion, Derrida also returns us to the phenomenon of sound (*Klang*, *Glas*). With the image of the tightrope still present in our minds, he compares the straight line of a suspended string to an oscillating or plucked line that, in its vibration, generates both curves and sounds. He writes:

> in the course of repeating the "birth of sound"... the act by which strings, pipes, bars vibrate is described as an "alternating passage from the straight line to the arc". In the process of subjectivizing idealization that the oscillation (*Erzittern*) and the vibration (*Schwingen*) punctuate, the difference between nature and spirit corresponds to the difference between what does not resonate starting from (it)self, the bodies (*Die Körper klingen noch nicht aus sich selbst*), and what resonates with (it)self. The history of *Klang* is what reappropriates itself up to resounding *with* Klang *itself*.[54]

Spirit comes into its own by way of internal self-production, what Hegel calls in the *Phenomenology* 'spiritual shape' (*geistliche Gestaltung*): into 'an inner (*Inneren*) that utters or expresses itself out of itself and in its own self; into *thought which begets itself*, which preserves its shape in harmony with itself and is a lucid, intelligible existence.'[55] This spirit, as *Geist*, is no longer merely the artisan or artificer, crafting things from the external world, nor even from the human bodily form, but the *artist* who creates from the freedom within. For Derrida, hearing accompanies this turn towards the interiority that is artistry. For hearing is the privileged sense, being more internal, more idealised, than all the others: the vibrations on the tympan convert sound to thought. Resonating within *produces*: external noise (*Geräusch*) transforms to internal sound (*Klang*), while internal sound transforms to the voice ('hearing is the receptive, conceiving voice'),[56] and voice transforms to language, to sense and to thought. The artistic spirit, then, is that which speaks a language through the inner freedom of its own internal shaping, 'the spontaneous outside production of an inner sense... [where] the production of self by self gives voice'.[57] While the artificer is caught in a negative interiority, since what s/he depicts *from* within is still the exteriority of itself, self-consciousness shaping the natural, the artist by contrast shapes *within*, self-consciousness shaping *itself*. The former is exemplified by the ancient Egyptians, with their sphinxes and Memnon statues, the latter by the ancient Greeks, with their drama. But the spirit here is still the spirit of religion, remember. Natural religion has given way to religion in the form of art, with its higher element of language. The work of art must speak, must give voice as the spirit (the *Geist*) within. So religious hymns, religious oracles, religious poetry, religious drama.

What makes drama in particular so *spiritual* is its interiorisation of the artistic voice: the art speaks, and speaks from within, but in the outer expression of an actor's voice and of the staged drama as spectacle. The *spirit* of this dynamic, whereby the interior is made exterior, and the exterior interior, the invisible *Geist* that operates on the threshold, like a transparent membrane in the ear, is also the work of the gap between the exteriority of art and the interiority of philosophy, between the two columns of *Glas*. The spiritual work of art, before it becomes revealed religion, is thus that moment when the god, as spiritual being, must give himself up to the blankness that is the void between total interiority and total exteriority. Hegel, in the *Phenomenology*, uses the language of kenosis to describe the Spirit who empties himself in 'complete corporeality'.[58] Derrida, at the very end of *Glas*, picks up on this self-emptying as he describes the Dionysian cult that gives way to tragic and comedic drama, and ultimately to the revealed religion of Christianity. Dionysus, who inspires frenzied stammering within his (female) followers, must 'pass into his contrary', must 'appease himself to exist', no longer the cultic figure or 'the sculptural column of divinity'.[59] But neither must he be pure aesthetic beauty of the human shape or form (as in the Greek athlete or gymnast). He must pass from noise to speech, from stammering to language, from mystery to *Logos*. On the one side, 'delirium, the self (*Selbst*) [that] loses consciousness'; on the other side the stage, where 'spirit is what is outside of itself'.[60] And at this point a revelation should occur: the revealing of absolute Spirit who shapes itself through and as self-consciousness, the Mediator stripped both of natural aspect and of abstraction, the Christ figure whose death is also a reconciliation of both sides.

But Derrida cuts off his text before any such revelation can occur (in Hegel). Instead, he ends with a Nietzschean connection: 'A time to perfect the resemblance of Dionysus and Christ.'[61] What can this mean? He suggests there remains a family resemblance to be drawn distinctly now between the Dionysian enthusiasm of the bacchante and the incarnated *Logos* as fully revealed in language. Yet what is this resemblance? We might answer: a speculative resemblance. For is not the kenotic death required by both in order to be rendered properly into the Hegelian *Geist* that very self-sundering of *Geist* itself, spirit outside itself – not just in art (as in the inspired drama, whether that of the Dionysian festivals in Athens or that of the passion at Calvary),[62] but now in philosophy (of Hegel) and in religion too? If language is to be the final reckoning of these movements as spiritual art, is not language itself implicated in this self-emptying that reconciles only through its absenting? So when Derrida says in the penultimate line that between these two figures, Dionysus and Christ, 'is elaborated in sum the origin of literature', he means to say that the writing of the spoken word (the dramatic voice, the *Logos*), as the exteriorisation of that inner world we associate with thought, relies upon a certain absence and absenting, a negation both of abstraction (divine realm, mystery, thinking, etc.) and of physical nature (the body, the flesh, the materials of art). The *sparagmos* of the cultic frenzy – the literal tearing apart of the Dionysian hero/victim – and the Passion of Christianity – the crucifixion of the hero/victim – are both ritualised deaths through which Spirit begets itself as self-consciously *human*. Literature can only arise in that self-consciousness: the exteriorisation of the inner voice. But it arises already ruined, split by the gap that separates its two sides. This is why the 'last' lines of *Glas* ('last' until, like its Joycean counterpart, it throws us back to the beginning) must speak ruinously: 'But it runs to its ruins {*perte*}, for it is counted without {*sans*}' (the left Hegelian column); 'Today, here, now, the debris of {*débris de*}' (right Genet column).[63] Where the sound of the tolling bell might resonate onward in these unfinished thoughts, the literary line cuts itself off abruptly – 'without...', 'debris of...'. One must either search in the blankness between, the religious spirit of the textual gap that one dreads, because one loses oneself, or return to the opening lines that had already signalled the 'remains', the remains of a philosophical and artistic text that, as the penultimate line of the Genet column tells us, 'republishes itself'.

Derrida's intense project here, in a text that affords no way to soften its intensity without losing the force of its prolificacy and creativity, is an implicit celebration of negation out-negating itself, and therefore out-negating Hegel. Art should give way to religion, which should give way in turn to philosophy. But here philosophy is implicated in the artistic ruin of abstraction, while art is implicated in the philosophic ruin of materiality. Religion sits both between and within each side, not to mediate happily through mutual reconciliation, but to ruin all the more, by means of its own kenosis of text, a sacred literature that remains absent, or remains present as marginal nothingness, blankness. *Aufhebung* in all its speculative spirituality. This is to say that *Glas*, contrary to all previous accounts, remains Derrida's first great religious text, that is, his first text about religion, and should give us a clue about his later supposed 'turn to religion'. As *Glas* itself informs us, there was never any 'turn to', but only a 'turning back upon'. Yet such return is possible only on the pivot of negation, which is why Derrida's later writing on religion has so often been construed, rightly or wrongly, within the parameters of a negative theology.

Ironically, Derrida never explicitly writes in any sustained manner about Hegel in his later work.[64] But then, we could say he is never not writing about Hegel. He writes out of an obliqueness, an ellipticity, which is true to the negating *Spirit* of Hegel, which can only emerge properly when it is outside itself, negating itself. But that negation, we have seen, is profoundly generative. It generates 'literature', for one. It generates the spiritual work of art as religion, for another. But more, it generates itself as *Glas*, the text that cannot be properly read except by negating the conventions of reading (philosophically, religiously, aesthetically). To remain true to this reading, one must constantly read elsewhere. If Derrida's Hegelian legacy passes anything on to us, it is precisely this elsewhere – elsewhere to Hegel, but also Hegel as elsewhere. We turn now to those who keep this elsewhere living here in our contemporary world.

CHAPTER 6

The Living of Hegel and Negation: Kristeva, Nancy, Agamben, Žižek, Malabou

It is impossible to read contemporary appropriations and figurations of Hegel's negation without going through the prism of the Derrida we have tried, in a dangerously foreshortened manner, to configure in the previous chapter. For as Derrida himself exemplified, any taking up of Hegel and his negation is always both a re-appropriation and a dis-appropriation, a re-figuration and a dis-figuration, any of which can be made only through another. This alter-ation might in fact be another Hegel, but now a Hegel as the *Aufhebung* to himself, which always manifests itself in another thinker. Such an *Aufhebung* – the cogenitivity within 'the *Aufhebung* of Hegel' – has a long nineteenth- and twentieth-century history (Kierkegaard, Marx, Feuerbach, Kojève, Bataille, Heidegger, Sartre, Hyppolite, Blanchot, Adorno, Derrida…), and it tells us something about the history of Hegel: history works by negation. If we have arrived now at a living history, a history made present in figures still living, this only says that the present negation is itself historical, and must be read through its history. Derrida is now a spectral figure in this history, a spirit who remains behind each of the thinkers we now (selectively and briefly) turn to, even if their own work is elsewhere to Derrida, or elsewhere to his Hegel.

JULIA KRISTEVA

In her early book *Revolution in Poetic Language,* whose English translation is an abridgement of her State Doctorate *La Révolution du langage poétique* presented in 1973, and published a year later in 1974, the same year as Derrida's *Glas*,[1] Julia Kristeva provides one of the most significant appropriations of Hegelian negation, with an elaborate translation of its profound disruptive nature as it is intimately tied to 'art'. What concerns her principally in this text is the process of signification, or the signifying *process*.

She begins the dissertation by drawing an idiosyncratic distinction between the semiotic and the symbolic in relation to language, linguistics and signification. Linguistic theory has traditionally understood language as an object, something formally and statically self-contained. One studied any given language and its linguistic components as if they were a given and self-autonomous whole. One did not study the *process* by which that language might have come about, subjectively and inter-subjectively. The terms semiotic and symbolic are meant then to relate language to 'external' factors of process beyond its static internal structures (syntax, grammar, etc.). The first, semiotic, pertains to psychic drives and motivations such as one might find in the Freudian unconsciousness, with their connection to psychoanalytic and psychosomatic discourses. The second, the symbolic, pertains to more pragmatic and semantic reductions of language to logical modal relations such as one might find in Chomskyan generative grammar or theories of enunciation, with their 'deep structures' and categories, and thus their connection to phenomenology, logic and even philosophy. Kristeva sets up these two 'modes' as fundamental to any signifying process, and will elaborate their polarities with other pairings such as subjectivity/sociality, or genotext/phenotext (language's underlying foundation/language's communicative performance).

We might immediately see Hegelian pairings analogous to these: faith/knowledge of the early Jena period, for example, or in the later Jena period, the *Phenomenology*'s force/understanding, or in the *Philosophy of Mind*, mind subjective/mind objective. Kristeva in fact appears to have very much in mind the force/understanding of the *Phenomenology*, since her opening epigram is from the *Phenomenology*'s Preface – 'What, therefore, is important in the *study of Science*, is that one should take on oneself the strenuous effort of the Notion [*Begriff*].'[2] A later chapter picks up directly Hegel's conception of Force (*Kraft*) as an active power that, on the one hand, expresses a dispersal of individual matter or appearances within phenomena, but that also, on the other hand, operates as a pure *Begriff* behind those phenomena ('the Notion of Force *qua* Force'),[3] or what Kristeva describes as something 'that will produce a *non-objective inner world*, a return of Forces as a Notion within Understanding'.[4]

However construed (and we'll return to the question of force below), Kristeva understands these two sides or modes, the semiotic and symbolic, in the manner of Hegel: that is, *dialectically*. Unconscious drives or motivations do not operate independently from their social and logical manifestations. Nor is the one simply an outward expression of the other. In the practice of signification, there is a constant movement or procession between the two, so that no one side dominates over the other, and indeed no one side can make sense or be understood without the other. The emphasis must be on the flow between them, a flow that does not so much culminate in a third term that is higher than the two sides, but in an excess that flows outside both.

In the first section of her book, Kristeva calls this excess, or this outflow, 'poetic language'. By this term she does not refer to a narrow expression of language that takes the form of poetry. Rather, she employs it, much as Bataille had done, to refer to a more disruptive sense of language that exceeds the formalities imposed upon it by linguistic and anthropological sciences, and domesticated by common usage. This kind of disruptive language is seen best in art, myth, religion and rites, not just in poetry in the technical sense. It disrupts both the semiotic and the symbolic, both the individual and the social structures under which the individual lives. Kristeva writes: 'Magic, shamanism, esoterism, the carnival, and "incomprehensible" poetry all underscore the limits of socially useful discourse and attest to what it represses: the *process* that exceeds the subject and his communicative structures.'[5] This process of excess she further terms *signifiance*, which not only goes beyond the two sides of signification (semiotic and symbolic), but in fact brings them into being. *Signifiance* is thus generative, or 'is precisely this unlimited and unbounded generating process, this unceasing operation of the drives toward, in and through language; toward, in, and through the exchange system and its protagonists – the subject and his institutions'.[6]

In the second section of this work, Kristeva adds to the definition of this excess: as poetic language, excess is also *negativity* (Hegel's *Negativität*). This negativity is 'both the cause and the organizing principle of the *process*' that stands behind excess, and indeed stands behind the coming into being (poetically) of language as excess. She distinguishes this notion of negativity from Hegel's *Nichts* and *Negation*, as Hegel elaborates that notion in the *Phenomenology*,[7] and emphasises its mediatory role (between being and nothingness, abstraction and concretion, self and other, subjectivity and objectivity, etc.). But when she draws out its motive power, in what the *Phenomenology*, summoning Aristotle, calls '*being-for-self* or pure negativity',[8] or what she herself calls a 'mobile law',[9] she construes the term in the same way we, in drawing primarily on the *Science of Logic*, have construed the term 'negation'.[10] Thus Kristeva calls negativity the 'logical functioning of the movement that produces the theses [of being and nothingness]', or more positively, 'the liquefying and dissolving agent that does not destroy but rather reactivates new organizations and, in that sense, affirms'.[11]

More importantly, this movement of negativity, as the logical expression of the process Kristeva is trying to describe in relation to language, nevertheless stands outside logic as we normally understand it, and even outside the logic of the Hegelian dialectic as it is normally rendered. Its movement does not remain bound to a 'triplicity', in which a third term stands as a culmination of two opposing forces, now equally surpassed and conjoined. Negativity, in rejecting this kind of logic, and indeed all logic of self-containment (autonomy), introduces a 'logic' of rejection that is excess (heteronomy), and in doing so Kristeva, following Hegel at the end of the *Greater Logic*, can call negativity the *fourth term* of the dialectic – that which exceeds even the Hegelian system, while at the same time being fundamental to it.[12] This fourth term disturbs the unity of Self and Being, and therefore of language itself.[13] As Kristeva writes, 'Hegelian negativity prevents the immobilization of the thetic, unsettles doxy, and lets in all the semiotic motility that prepares and exceeds it.' It is therefore a moment of rupture, a 'productive dissolution'.[14]

We can see here the connections Kristeva makes with the generative aspects of negation we have been exposing throughout our discussion. Language, she suggests, as rooted in both semiotic and symbolic forces, comes about through the productive ruptures that are internal both to the deep structures of language as a system and to the social matrices that govern the enacting of language in any one form. We do not simply inherit our language from the standing wells of tradition; we generate language from undercurrents that we access by rupturing the ground of our subjectivity, and that explode onto the surface in an excess we can only contain by imposing some form of fixity (position, as grammar; proposition, as judgement or thesis; disposition, as habit – that is, the '*thetic*', in Kristevan terms). That excess is 'poetic', not in Bataille's senses of, say, sex and death, but in a sense that captures the creative, productive side of its generation, as well as the heterodox nature of its signification.[15] It is also, by that virtue, revolutionary, because it necessarily disrupts traditional positions, or overturns the status quo. Thus we might speak of negativity as a *poetic force*.

What most interests Kristeva about this force is therefore its combination of formative and disformative movements, of unity and scission, of independence and subjugation, of harmony and discord, of fecundity and ruination, of creativity and destruction. She finds these *coincidentia oppositora* already at work in the Force (*Kraft*) expounded in the third chapter of Hegel's *Phenomenology*, 'Force and Understanding: Appearance and the Supersensible World', where Force is a power that, in consciousness's apprehension of an object, both unifies and diversifies, through an *Aufhebung* that is sustained in the *Begriff*. (And the very act of this sustaining is what Hegel terms Understanding (*Verstand*)).[16]

Negativity, as poetic force, ultimately carries one beyond the two sides, and therefore beyond both Self and Being. It is thus 'trans-subjective, trans-ideal, and trans-symbolic', or, 'Hegelian negativity, aiming for a place transversal to the *Verstand*, completely disrupts its position (*stand*) and points toward the space where its production is put into practice.'[17]

Such a place, as poetic, is very often literary, and in the final sections of the abridged English translation of the dissertation, though throughout the original French full version, she addresses the text as practice, specifically in relation to the writings of the French poets Lautréamont and Mallarmé. Here the text is 'a practice that pulverizes unity, making it a process that posits and displaces theses'.[18] This pulverisation is consistent with the *process* that she began with involving the semiotic and the symbolic, for language, or signifying practice, finds its most potent, its most productive manifestation in such extreme texts as modern(ist) or avant-garde poetry, where the excessive nature of the process is put most forcefully into play. Not only is language pushed to an extreme, or an excess, along with the form which it takes as text, but also the subjectivity that is supposed first to issue that language, and then make a distinction between its content and its form – this too is pushed to an excess, beyond itself. The subject itself 'is an excess: never one, always already divided… [and] divided in such a way that the subjective "unity" in question is expended, expending, irreducible to knowledge'.[19] Language, subjectivity, form, content – all receive the same kind of pulverisation in the process that is a 'productive dissolution', a pulverisation that is always *in process* because it is itself process, the process of 'affirmative negativity'.[20] And by the end, Kristeva will say that this kind of practice encompasses the ethical, by the very fact that it takes us away from a fixation upon the transcendental Ego, the mastering One, or the narcissistic Subject, and divests us of our distanciated and distanciating abstractions. Here, art becomes critical.

> The text, in its signifying disposition and its signification, is a practice assuming all positivity in order to negativize it and thereby make visible the *process* underlying it. It can be considered, precisely, as that which carries out the ethical imperative. Given this insight, one cannot ask that "art" – the text – emit a message which would be considered "positive"… the text fulfils its ethical function only when it pluralizes, pulverizes.[21]

Kristeva concludes by contrasting this with Hegel's idealist position, at least as it emerges in the *Aesthetics*, where art functions to liberate us from the passions or 'the power of the sensuousness', and therefore must give way in the end to philosophy. The practice of art will always disturb philosophy, even revolutionise philosophy, and any ethics it may demand, because art, as party to the fourth 'term' (though art, as Kristeva is aware, never operates in such 'terms'), negativises the notion of philosophy as the philosophy of Notion (*Begriff*). This does not mean it dispenses with the Notion; rather, it means it pulverises it into a place where it can be made anew. It returns it, we might say, to its origins, its beginning.

Kristeva had already broached the subject of language as negativity in an earlier essay on Roland Barthes entitled 'How Does One Speak to Literature' (1971). There her question was of a similar kind, a dress rehearsal of her dissertation: 'How does there emerge, through its practical experience, a negativity germane to the subject as well as to history, capable of clearing away ideologies and even "natural" languages in order to formulate new signifying devices?'[22] Barthes, who would be an examiner for the subsequent dissertation, was here the focus of her attention because Barthes, for Kristeva, was the first to understand language and writing as inherent negativity: 'contestation, rupture, flight, and irony'.[23] Kristeva wants to show that Barthes' concept of 'writing' operates between the deep interiority of subjectivity and the substantial exteriority of objectivity. But the 'between' here does not unify the two, and least of all in a homogeneous and self-contained language. Rather, it allows each to be 'dialectically constituted' in relation to the other: 'it brings one back to the other, neither subjective individuality or exterior objectivity, it is the very principle of Hegelian "self-movement" and offers the very element of law.'[24] But this law is not the Law that Kristeva found emerging from Hegel's Understanding in the *Phenomenology*, where negation, acting between subjectivity and objectivity, between sensible and supersensible, becomes in fact the very difference between these two opposing realms, and thus a *universal difference* lying in the very heart of the appearance (or appearing) of any one thing, a universality expressed in a *law*.[25] For Kristeva, such a law is still predicated upon a unified subject of understanding (and a unified understanding of subject). Its result is a 'homonomic' sense of meaning. Writing, on the other hand, establishes a 'divided subject, even a pluralized subject, that occupies, not a place of enunciation, but permutable, multiple and mobile places; thus it brings together in a *heteronomous* space the naming of phenomena (their entry into symbolic law) and the negation of these names (phonetic, semantic, and syntactic shattering).'[26] This shattering, the 'pulverizing' of her dissertation above, leads not to no laws (*anomia*), but to multiple laws (*polynomia*), an excess beyond both subjectivity and objectivity, which characterises the fundamental function of writing. And this excess, according to Kristeva here, exceeds even Hegelian 'self-movement'. (But it can only do so if that movement must inevitably, and of its own accord, lead to a higher law of unity, a universal unity, and as we will see below, this is contested. Žižek, for example, who also picks up on the 'fourth term', will argue that the subject of Understanding is always already split within its own self, is already pulverised by a polynomial negation that constitutes the subject even before any objective negation (including the subject as its own objectification) is directed against it.)

Kristeva will incorporate these ideas of negativity – negation as the dissolution of wholeness, but also the opening up of polyvalent space – in writings subsequent to her dissertation as well. Her concept of abjection in *Powers of Horror*, for example, carries the same genetics as that of the earlier negativity, as it is linked with various literary figures: Dostoyevsky, Proust, Joyce, Borges, Artaud and Céline, for example.[27] But the context out of which it arises is more psychoanalytical and sociological (Freud and Bataille, e.g.) than it is linguistic and philosophical, and hence Hegelian negation never appears with quite the same Hegelian force as in *Revolution in Poetic Language*.

What we can conclude is that, as a fourth 'term', negation (or negativity) is for Kristeva part of the disruptive forces of philosophical discourse, and Hegelian discourse especially, even as it is not merely parasitic upon, but wholly emergent from the Hegelian scheme. Its emergence, as procreative, is therefore also its break from its source, its excess. Not three, but *at least* four. That this break is deemed *poetic* is precisely where negation comes into its own. It is in-itself and for-itself only when it is productive of its own dissolution, yet that dissolution is its own prolific and prolix beginning, through what Kristeva calls a 'heterogeneous contradiction',[28] that is, a negation that keeps on giving itself with creative abundance.

JEAN-LUC NANCY

Perhaps no other living figure has understood and elaborated Hegelian negation as a 'fourth term' more than Kristeva's compatriot Jean-Luc Nancy. Even more than Kristeva, Nancy's engagement with Hegel emerges from within the penumbra of Derrida. But like Kristeva, Nancy follows Derrida in foregrounding the creative possibilities of negation. In standing outside the system, in order to underwrite the system, Hegelian negation originates – not merely the system itself, but more significantly the internal dynamic of the system that operates against its own structural integrity and comprehensive unity. As with Kristeva, this contradiction is plural, multiple, polyvalent. *At least the fourth*. Negation clears the ground of *all* singularity (in all its contradiction). Or, closer to Nancy's own language, negation 'makes us available' for what is coming, as it makes the coming available (in us, as us, for us, against us).

In the same year that Kristeva wrote her dissertation, and a year before Derrida published *Glas*, Nancy published *The Speculative Remark*, an extended analysis of what he called, in the subtitle, 'one of Hegel's bons mots'.[29] The 'remark' in the main title (Hegel's *Anmerkung*) becomes a 'bon mot', a good, right, correct word that is both supplement to but also centrally significant in the main argument. The additional becomes principle, and therefore a clever little trick of word and thought: what goes beyond also comes back to its centre. This 'bon mot', as a word, is not yet 'negation', 'negativity' or 'the negative'. But it is intimately related to this family of negatives, since it is, we might say, the very enactment of negation, or the word for that enactment. This word is precisely *Aufhebung*. And its enactment is the *sublation of the negative* in the fullness of the cogenitive, that is, the sublation of the sublation (of the sublation... and so on) of the negative, and the negation of the negation (of the negation... and so on) of sublation. This enactment becomes Nancy's text itself, for the remark with which he begins is Hegel's own *Anmerkung* in the *Science of Logic* on the term *Aufhebung* ('Remark: The Expression "To Sublate"'), so that *The Speculative Remark* becomes Nancy's own speculative *Anmerkung* on Hegel's *Anmerkung*, and ultimately a sublation of that expression to sublate as Hegel himself speculates upon it. We clearly see a Derridean strategy here.[30] But the play on the *bon mot* (in German, *Witz*, a witticism or a joke) will not take a typographical form; it will be internal to the speculative reasoning itself.

The sublation of negation, as it unfolds in Nancy's text, follows remarkably along the same lines as Kristeva's simultaneous elaboration of negativity as process. Let us recall that Hegel's *Anmerkung* on *Aufhebung* (and its verbal form *aufheben*) in the *Greater Logic* is an additional remark to the short section on the 'Sublation of Becoming' (*Aufhebung des Werdens*) that concludes the first chapter on Being. The sublation that Becoming enacts is the sublation of Being and Nothing, which, when sublated, leaves Becoming. Becoming is therefore always being enacted between these two poles, and so Hegel calls it an 'unstable unrest' (*haltungslose Unruhe*),[31] whose moments are forever repeating themselves. But in order for any determinate Being to arise, Becoming must sublate not just Being and Nothing but its own self. In doing so, it stabilises into rest, or into a stable oneness that is determinate Being (*Dasein*). The sublation of Becoming is thus likewise cogenitive: it acts both objectively and subjectively. In the *Anmerkung*, Hegel qualifies this sublating process ('one of the most important notions in philosophy') as always preserving that from which it originates: 'Nothing is *immediate*; what is sublated, on the other hand, is the result of *mediation*; it is a non-being but as a *result* which had its origin in being. It still has, therefore, *in itself* the *determinateness from which it originates*.'³² But Nancy remarks that such a process of preservation acts upon the process itself. What, we might ask, originates the *Aufhebung*? What is *its* own becoming? It is not Becoming as it stands in the triplicity of the dialectic between Being and Nothing. For that Becoming, by the end, is sublated into a stability, a rest (determinate *Dasein*, the subject of Chapter 2). Rather, it is the becoming that *is* sublation, *Aufhebung*, whose moment is a *ceaseless* unrest. *Aufhebung*, Nancy says, is therefore always *sich Aufhebung*, self-sublation,³³ which is to say that it is not determined by something outside it, not even by Becoming, but is its own becoming that in turn determines Becoming, 'that must have taken hold of [B]ecoming and resolved it'.³⁴ It exceeds the triadic movement, therefore, by evading its schema, by vanishing within the text through a certain Derridean sliding, vanishing within its own self to preserve itself, approaching itself by means of distancing itself.³⁵ It is in this sense that Nancy writes: '*Aufheben* is indeed *the* word of Hegelian discourse, the *right* word [bon mot] for speculative thought, its password.'³⁶

But as a password, as a *bon mot*, it is multiple, what Kristeva would call heterogeneous. In erasing itself, in causing its own vanishing, it brings us back to its beginning, much as we began our text – returning anew to and with Hegel – by forcing us into yet another pass. The *Aufhebung*, as the negation of itself, is a proliferation of beginnings. 'That is to say, to begin (or to begin again), that it gives to a multiplication of texts.'³⁷ So we have Hegel's text of the *Greater Logic*, both the original (1812) and its revision (1831), the latter of which Hegel says required 'a fresh undertaking, one that had to be started right from the beginning', one that demanded revision 'seven and seventy times'.³⁸ We have Hegel's Remarks on those texts (Remarks which themselves have been revised). We have another version altogether, *The Encyclopaedia Logic* (*Lesser Logic*), itself in three editions (1817, 1827, 1830), with both Remarks and the posthumous Additions (*Zusätze*) added by von Henning in 1840. We have Nancy's remarks on those texts (alongside Derrida's, Kristeva's and a host that preceded them), and even Nancy's later remark (1999) on his own text.³⁹ And we now have our own remarks on Nancy's texts. And on it goes. This proliferation is not merely the begetting of commentary in the exegetically profuse fashion of midrash, nor some unending Borgesian textual labyrinth that characterises all language and writing, even if it might lead to elements of both of these. It is more the primordial nature of *Aufhebung* itself as an unceasing negative motility, a negation that is wholly circular in its cogenitive attachment to itself, by which it produces its own self (and non-self) from within.

Nancy calls the function of such attachment 'speculative', after Hegel, meaning that the work of *aufheben*, of simultaneous suppressing and preserving, and of deep reflexivity, is not grounded in any logical or grammatical or syntactical or semantic totality, structure or principle. And by that virtue, as Kristeva argued in her own way, it does not lead to some higher positive state. Speculative meaning, which is both the meaning and the process of *aufheben*, is ultimately *without meaning*. The meaning of the word *aufheben* is precisely that it has no meaning – and this is not, Nancy tells us, in the sense of harbouring the nonsensical, nor even in the sense of sublating meaning, but rather in the sense of *exhausting* meaning.[40] Meaning is engulfed and consumed and left nihilated by *aufheben*. But not because *aufheben* is a destructive path. Rather, because it operates in some sense *outside* meaning, or *in excess* of meaning, not as its condition, but as an opening that has yet nothing in it, nothing to it, except the wellspring of some future possibility. If it exhausts, it also inspires.

But to exhaust and inspire both at the same time leaves the breath suspended. Near the end of his text Nancy describes (after the fashion of Nietzsche) this strange paradox as a tone without an accent. 'Sublation,' he writes, 'is not a rhythm, "a swinging back and forth" between two terms or two poles in turn accentuated.' Rather, if it is anything at all, it is a *'meter without poetry'*.[41] It exists – it can be experienced or sensed – but as a 'word' it slips and vanishes amid the syntax and meaning, and thus cannot be distinguished as a 'word' in any proper sense. It negates its own logocentricity, or its own logical operation. And in this way Nancy links negation, now fully as speculative *Aufhebung*, and following Hegel himself, to a productive imagination (*Einbildungskraft*).[42] But this imagination would not be the kind we normally associate with this term, for it is neither the productivity nor the imagination of an artist creator who crafts something that can be placed before us for the purposes of *aesthesis*, the sensory beholding of an object (in beauty or admiration). It is, instead, a productive imagination that sets into play, or rather is the eternal playing of, a tremor, a motion, from which something – poetry, say – might arise. This playing is the play of negation.

Twenty-four years later, Nancy returns to Hegel anew, in *Hegel: the Restlessness of the Negative (Hegel: L'inquiétude de négatif*, 1997). The motion of negation's play, the metre of negation, is now made emphatically restless. We have seen the German noun *Unruhe* above in relation to negation as unstable within the *Greater Logic*. But Hegel employs this term *Unruhe* frequently throughout his corpus to characterise negation, even as early as the *Phenomenology*, where, in the concluding pages, he speaks of the self-alienating Self as 'its own restless process [*Unruhe*] of superseding [*aufheben*] itself, or negativity [*Negativität*]'.[43] The restless nature of negation thus inheres in the notion of *Aufhebung*. It agitates, it does not stay still. It persists like a current, but remains as invisible. Thought, *philosophical* thought, and maybe even theological thought, is meant to fix this restlessness, to stabilise it in certainties, such as those of logic or doctrine. But Hegelian thought – the *Aufhebung*, the speculative meaning – is always an undercurrent, always ripping our feet out from under us. Nancy declares: 'what thought is certain of is its restlessness, just as what unsettles it is its certainty.'[44]

In a remark to the earlier text *The Speculative Remark*, written two years after the emergence of *Hegel: the Restlessness of the Negative*, a remark he entitled 'The Speculative Unrest', Nancy claims, amid a certain apology for the 'stylistic and lexical complication' that was fashionable in the early 1970s in French thinking (as evidenced in *Revolution in Poetics* and *Glas*, its contemporaries), that restlessness (*inquiétude*) was already operating as a silent motif within his speculations on *Aufhebung*.[45] 'That is to say that the rapid and incessant movement that goes simultaneously for and against the "proposition" in order to achieve "speculation"… disrupts the essential mainspring of the system, and therefore makes the system restless, makes it anxious, drives it wild, prevents absolute knowledge from absolutising itself', and that it is precisely this unrest 'that gives Hegel's text its greatness and its strength'.[46] In *Restlessness of the Negative*, Nancy will explore this strength as the 'infinite work of negativity', a work that does not merely seek or find itself, but that 'effectuates itself' as the 'living restlessness of its own concrete effectivity'.[47]

The book, in dealing with a particular Hegelian theme throughout each of its eleven chapters, might be said to follow loosely the progression intended in the overall philosophical system that the tripartite *Encyclopaedia* attempts to expound, starting with Logic (Nancy's chapters 'Restlessness', 'Becoming', 'Penetration', 'Logic'), moving through Nature and History ('Present', 'Manifestation'), and culminating in *Geist* ('Trembling', 'Sense', 'Desire', 'Freedom' and 'We'). Each is shot through with the negative. Nancy's narrative begins with the stirrings of the negative as *Unruhe*, which comes about through the productive reflexivity of relating oneself to itself, a relation that is inherently negative, since it relates only to its own internalisation (its own immanence, we might say), at the expense of that which is beyond it. It continues with becoming, now with a circularity that begins to mark the Hegelian logical movement:

> Hegel neither begins nor ends; he is the first philosopher for whom there is, explicitly, neither beginning nor end, but only the full and complete actuality of the infinite that traverses, works, and transforms the finite. Which means: negativity, hollow, gap, the difference of being that relates to itself through this very difference, and which *is* thus, in all its essence and all its energy, the infinite act of relating itself to itself, and thus the power of the negative.[48]

This power then becomes the power of penetration, since it penetrates its inner self as an other, its negative relation. And this penetration, as othering, or separation, is at the very 'basis' of Logic, for logical thought is not something that comes to us from without, as if given to us, but is derived from its own internal powers: 'Logic is thus, from its most elementary stage onward, from its first and poorest abstraction, a tearing of identity out of itself, its dislocation, and its alteration.' It is also, by this token, the negative movement of self-identification or self-unification, whose productive reflexivity brings into being its content. Thus Nancy can write, '*Logos* designates the "making" of every "given" – that is to say, its "giving" and, more precisely, its "giving of itself": thus *logos* designates the identical not as a substance but as act.'[49] It is only by way of this kind of logic (a logic that confounds most conventional logicians) that we can say that the negative, as the act of negation, is an act of its own making. Or that Hegel, and all his Hegelian offspring (which more often means Hegelian*isms*), unmakes himself and unmake themselves as an act, necessarily, of their own doing, and as a measure of their self-enervating strength.

How does one make present this paradox of negation? Or why does its restlessness remain present today, at this very moment? It is because, Nancy tells us, the present is only possible through a negative self-relation, which is to say through a relation to its history that must be made anew here and now – a making which is precisely its history, the history of its making (cogenitively). The world does not arise and progress out of its past towards some advancement in the future, which is beyond itself. The world is of its own immediate making: its finitude is always resident within its infinitude. The presence of the world is precisely this productive dissolution of itself, this separating out of the one (past, present or future) from its opposite, in order to bring them into a relation through that separation. It is thus a manifestation of itself through the negative, a singular manifestation of the world we know as phenomena: 'The singularity of manifestation, or of the world: it is that singularity manifests itself to nothing other than itself, or to nothing. Manifestation surges up out of nothing, into nothing.'[50] This nothing is its self, and thus philosophy is 'the self-knowing of negativity even as it is the knowing of the negativity of self'.[51]

So the stability that we normally associate with the presentation of the phenomena, the world that is there in its present concreteness and compactness, is disintegrated or dissolved by the penetrative movement that our own self-made-knowing makes manifest. This means that the self, either as world or as individual, is always undergoing a kinetic disturbance, what Nancy calls a trembling (*tremblement*). Like the *Aufhebung*, this trembling agitates us from within our own selves, and not from some external source. We tremble before our own making, and therefore we tremble *as* our own making, and unmaking.

The trembling and the restlessness are obviously intimately related. 'The subject is the experience of the power of division, of ex-position or abandonment of self.'⁵² We cannot penetrate any one thing without the self-divisive motivation that is trembling.

But how do we *make sense* of this motivation? 'Sense' throughout much of Nancy's work is always a fully laden term. As in the semantics of English, 'sense' in both German and French (*Sinn*, *sens*) can go in opposite directions: towards the concrete and empirical, as in the sensations, or the attendant faculties by which we can experience sensations; or towards the meaning and logic by which a word or an event might become understandable to us. It is thus at the heart of the Hegelian dialectic, since it carries both the physical and the ideal in a simultaneity of function. To maintain this simultaneity, it must remain in constant motion, as it vacillates between nothing, becoming and manifesting. This means 'that it is incessant movement and activity: as much the perpetual movement of signification in language as the movement in history in which nature and man never cease passing – in the double sense of being-in-passing, and passing away – and as the movement of acting, of human operation and conduct, which have to free, always anew, the truth of sense for itself.'⁵³ Sense thus involves the subject's own subjection to its lack of fixity, to being thrust out of itself and its own determination by its own powers. To appropriate this subjection, that is, to sublate the sense of the subject in both its directions (the sensing of the self as the meaning of the self, and the sense of the subject as both its own subjectivity and the subjection of that subjectivity to something other than itself) involves the very negative impetus that was unearthed in *Aufhebung*. 'The restlessness of the negative is the agitation, the tension, the pain and the joy of this appropriation.'⁵⁴

In desire, this double condition comes about through self-consciousness, which can be conscious of itself only in the consciousness of its otherness. For desire is the desire for this otherness, becoming other than what one is at present. There is no Freudian apparatus here, as in Kristeva, governing the nature of this desire. Desire is the *becoming other*; it is not a drive towards something or someone outside oneself that one does not possess, but the desire, the knowing impetus, *to be other*. Desire does not effect a relation between two determinate subjects (or objects). It effects their fusion in self-relation: both the concrete singularity of the subject in its subjectivity, and the negativity of that subject abandoning its concrete identity as subjectivity. This is what Nancy means when he claims, 'One must think concrete negativity.'⁵⁵ To make the negative concrete, one must combine both the singularity and its abandonment. Only in such negativity is freedom available. For freedom is not a liberation of the self that maintains its own desires, but the taking leave of that self, by means of desire, towards a freedom *as the other*. Freedom is not an inherent state, or even a right. Freedom is not a free self, in the usual sense that this is understood today – an autonomous self able to direct its life in the unencumbered liberty of its own decisions. Such a self, Nancy claims, is in fact the negation of freedom. 'Freedom, to the contrary,' he posits, 'is the negation of this negation, or negativity for itself.' This 'negation of negation' is the negation of lack that in desire is simply the lack of another thing outside me, the lack that constitutes me through the non-possession of an external object. One negates this lack as, we might say, lacking the other within. Negativity *for itself*, 'freedom', gives our own self to its own internal otherness. It is a freedom *with* that otherness, and it is a freedom *as* that otherness. That is why the 'journey' (circular as it is) must 'end' with 'we'. The singular will be made plural. And the plurality will be known, not as absolute knowledge, but as absolute *knowing*, an ongoing act of 'knowing in restlessness, knowing without rest',[56] a ceaseless knowing that moves in a circle, in order to trace out the zero of its negativity. This is a self-knowing that, like Kristeva's abjection, is neither objectivity nor subjectivity, but rather, as with Levinas, what Nancy calls 'our just-between-us [*entre-nous*], the just-between-us of our manifestations, our becoming, our desire'.[57]

So what Nancy fashions here is a rewriting of Hegel's own symphonic structure of Self and Spirit coming to absolute knowledge of itself (Self as Spirit), now with the murmur of negativity amplified, and with its divisive production placed to the fore. The restlessness never ceases, from the opening to the close; it propels the narrative forward. But that propulsion is also an internal dirempting, and Nancy concludes by quoting Hegel from the Preface of the *Phenomenology*: 'We never stop losing the "fixity of the self-positing".'[58] We lose it in *aufheben*, the speculative metre that furnishes a certain rhythm, but never one that beats between two distinct poles. The rhythm operates underneath the articulation, even in the absence of articulation: a metre without poetry, unfixing the words even before they are committed or enunciated, as with Kristeva's process. But nevertheless that metre also allows the words to take shape, to be placed anew, in the paradox of its *unruhig* sublation.

If *Aufhebung* is a metre without poetry, then poetry (good poetry) is forever aware of that metre, which dislodges its own feet, which loosens its very footing. This is why in the text Nancy can say that what is posited in art cannot be a matter merely of representing sense: we must, rather, enter art's 'movement and penetrate its act', and this will always mean we must be caught up in its negating movement, in the very act of negation as pro-ductive, a restless leading forth, which is both a deriving and a driving.[59]

The implications of negation's restlessness for art have been part of Nancy's corpus since the writing of *The Speculative Remark*, his first major text. They can be found on various registers in, for example, his co-authored book with

Philippe Lacoue-LaBarthe, *The Literary Absolute: The Theory of Literature in German Romanticism* (orig. 1978),[60] in *The Sense of the World* (orig. 1993),[61] in *The Muses* (orig. 1994), in *Listening* (orig. 2002),[62] in *The Ground of the Image* (orig. 2003),[63] and in *Multiple Arts: The Muses II* (2006).[64] These texts, and many others yet to be translated into English, cover a range of artistic forms, from music to painting to film to literature. In an indicative essay that opens *The Muses*, Nancy, in asking 'why are there several arts and not just one?', returns us to the notion of vanishing that was coupled with Hegel's *aufheben* above. The moment 'Art' is manifested as a singular work of art, he claims, it is no longer a unity, not even an instance of the 'arts', nor even its own self-contained singularity, but a multiplicity or plurality of material being.

> As soon as it takes place, 'art' vanishes; it is *an* art, the latter is *a* work, which is *a* style, a manner, a mode or resonance with other sensuous registers, a rhythmic reference back through indefinite networks. In a certain very precise sense, art itself is in essence *nonapparent* and/or disappearing. It even disappears twice: its unity syncopates itself in material plurality. (The moment of the Kantian sublime or that of the Hegelian dissolution is always present, at work *in* aesthetic 'immanence' itself.)[65]

We will return below to the question of art and its relation to religion and philosophy, but here we can see that Nancy's contribution to the question of Hegel's negation is that in its productive dissolution, in its generative diremption, in its effectuating separation, negation always moves us away from unity, and especially the absolute unity as totality that has characterised so many Hegelian readings since Kierkegaard. Negation, as the engine of the system, precisely undoes the system, never allows it to properly systematise, forever agitates it in a restlessness that is also an origination, breaking it down to build it up again. Even 'art' is driven by this restlessness, for in the end, that is to say, at its beginning, art in its plurality *is* this negation. It is unity being disrupted and fractured by plurality. It is art's own self being riven by its otherness. 'The "end of art",' Nancy writes, with the famous opening of the *Aesthetics* in view, 'is always the beginning of its plurality'.[66]

GIORGIO AGAMBEN

We find these meditations on art and its vanishing, or on the generative 'negation of art', also in the work of Giorgio Agamben. Two works stand out in this relation, both again from the 1970s. At the end of that decade, Agamben embarked on a series of discussions around the question of negativity in language, using both Hegel and Heidegger as a departure point. The resulting text was *Language and Death: The Place of Negativity* (orig. 1982). The questions raised there are very akin to the questions we have been exploring in Derrida, Kristeva and Nancy: the expression of negation and negativity as a power to speak, both in terms of how we are driven to express negation and, more significantly, how negation is driven to express itself. In the second work, which is in fact an earlier work, *The Man Without Content* (orig. 1970, Agamben's first book, but republished in Italy in 1994), that power is explored specifically in terms of art, using Hegel's conception of art, and once again focussing on the diremptive nature that inheres in this power, though now with different implications than previously seen. Let us deal with the former text first.

The 'place' in the subtitle of *Language and Death* is more than simply a generalised role negativity plays within the arenas of our speech, and in our condition of mortality. The placing of negativity comes, Agamben argues, at the very nexus of Being's locatedness in this world, as signified by the 'Da' (here, or there, in the ostensible terms of taking up ground at a particular place) in Heidegger's famous utilisation of the term 'Dasein', and by the 'Diese' (this) of 'Diese-nehmen' (lit. 'This-taking') of Hegel's *Phenomenology*, whose first chapter is 'Sense-certainty: Or the "This" [*Diese*] and "Meaning" [*Meinen*]'.[67] We will not pursue Agamben's elaboration of Heidegger, except to point out that Dasein's comportment towards the world, signified by the 'Da', is constituted by an existential comportment towards its ownmost death, so that negativity is at the very core of Being's placement, and that this constitution is precisely what is carried over from Being (*Sein*) to Determinate Being (*Dasein*) in Hegel's *Science of Logic*: 'Determinateness is negation posited as affirmative'.[68]

But it is not to the *Phenomenology* that Agamben turns in the first instance of his discussion, but to a poem written by the early Hegel in 1796 to his roommate of several years earlier, Friedrich Hölderlin. The poem was entitled 'Eleusis', in reference to the Eleusinian mysteries, and dealt with the ineffability at the heart of the mysteries, and their retreat from access and view in modernity. The young Hegel 'portrays himself here as the guardian of Eleusinian silence', a guardianship, significantly, that takes the form of poetry.[69] But the Hegel of the *Phenomenology*, a decade later, does not abandon this role, suggests Agamben. He simply alters his mode of guardianship. For the 'This' (*Diese*) of the first chapter on self-certainty, the 'This' that tries to mark out the place of the object before the senses, prior to that object being granted meaning, is no less shrouded in mystery, since language falters when it tries to articulate the sense-certainty in its place. 'Any attempt to express sense-certainty signifies, for Hegel, to experience the impossibility of saying what one means', writes Agamben. This is because the 'This' cannot capture the singular essence of a thing in its particularity, but only a 'universal This', which necessarily distances itself or abstracts from the sensuousness of the thing (a fact of which Nietzsche was all too aware). In Hegel's own terms, 'The Now is indicated, *this* Now. "Now"; it has already ceased to be in the act of indicating it. The Now that *is*, is another Now than the one indicated, and we see that the Now is just this: to be no more just when it is.'[70] To try to grasp 'This', to grasp it by enunciating it, and then giving it meaning, is thus to invoke negation, and a dialectical process between negation and mediation. At this point in the *Phenomenology* Hegel returns to the Eleusinian mysteries to exemplify his point, for the initiate's act of eating bread and drinking wine in the mysteries (homologous, of course, to the Eucharistic rite) is placed in the stead of the ineffable, so that in some sense these acts, these gestures, these rituals, guard that ineffable at the same moment as they manifest themselves. Their very concretisation and placement – '*this*' – puts a protective covering over the mystery, preserving it, but preserving it in its nothingness. By making the 'This' meaningful, by giving it a concrete location and situation, the negativity behind it is held in place. 'The content of the Eleusinian mystery is nothing more than this,' Agamben summarises: 'experiencing the negativity that is always already inherent in any meaning, in any *Meinung* of sense certainty.'[71] Though he does not draw the connection himself, Agamben might well have said: the young Hegel's poem is a form of ritualised protection of negativity in the same way as the eating of bread and drinking of wine is a gesture to the unspeakable, just as much as the *Phenomenology* is a making transparent of that protection, a transparency which of course is a mediation.

Agamben's gambit – to start with a poem by Hegel – is of more premonitory significance than even Agamben implies. The place of negativity may not, after all, begin at the philosophical point of a phenomenology or a science, even in Hegel's own *oeuvre*, but at the creative point where its mysterious placing emerges – unspoken – from the ruined sanctuary of the poetic act. 'And so you did not live on their lips', the poem states of the mysteries; 'Their life honored you. And you live still in their acts.' And thus, 'I introduce you as the soul of their acts!'[72] Hegel the poet introduces the ineffability of the act of the mysteries and their negation. It is not first the philosopher's prerogative, even if Hegel the philosopher might return to those mysteries at the outset of his great project, and indeed at its end.[73] Shakespeare perhaps sensed this same prioritisation: in *The Tempest*, as the drama runs to its conclusion, it is not Prospero's books that remind him of the negativity that is Caliban's machinations, but the masque he has presented, the theatrical performance invoking the Mysteries themselves.[74]

In his discussions to follow, Agamben turns to the phenomenon of the voice, and in tracing from Aristotle onwards a history of understanding the *phoné* or *vox* as something separate from language per se, Agamben follows a similar pulsation to what Nancy had proposed as a 'meter without poetry'. In the voice, before articulation or enunciation, there is present 'a pure intention to signify, as pure meaning, in which something is given to be understood before a determinate event of meaning is produced'.[75] But this presence as intention, the ostensible sound in something like a 'there' or a 'this', emits more than just a 'mere sonorous flux'; and yet it does not yet emit discourse, a matrix of grammar, meaning and logic that constitutes any given language. This in-between presence, what Nancy might have called 'our just-between-us [*entre-nous*]', Agamben calls 'Voice', which has gone beyond the 'voice' of sonic utterance, but has not yet reached the 'voice' of meaningful discourse. It therefore arises in a *negative* space, between states, so that 'that which articulates the human voice in language is a pure negativity', one which has an essential relationship to death.[76] Like Nancy's metre, this 'Voice', even if it utters 'I' (or, especially when it utters 'I') becomes the *Aufhebung* of the sound of breath, the sounded note of Self not merely as the harbinger of one's own negation of breath but that negation enacted. For Agamben the Voice indicates a *taking place* of negativity, a situating of that negativity before our language actually unfolds in its *Logos*, so that when we do finally speak the Logos, we 'speak with the voice we lack'.[77]

It is not surprising, then, that Agamben should end his contemplations with a return to the poet. For the metre he has been trying to elucidate between language and death is the metre that stands, first and foremost, before poetry, in every sense of this 'before'. So he invokes, among others, Sophocles, for whom poetry was still a living enactment, and quotes the famous words of the Chorus at the end of *Oedipus at Colonus*, perhaps the greatest ancient play of the ineffable mystery, even more than *The Bacchae* itself:

> Not being born overcomes all language; but, having come into the light, the best thing is to return as soon as possible whence one came.[78]

Agamben adds:

> Philosophy, in its search for another voice and another death, is presented, precisely, as both a return to and a surpassing of tragic knowledge [but this is nothing other than the *Aufhebung* at work

yet again]; it seeks to grant a voice to the silent experience of the tragic hero and to constitute this voice as a foundation for man's proper dimension.[79]

If the *Aufhebung*, as a metre without poetry, is at work in philosophy, how much more in the tragic voice itself, whose metre *is* poetry, a poetry that, at its best, strips itself from its own metre, as the Chorus suggests above? To the question of art's role in this stripping, we must move back, in a certain return to whence we came, to Agamben's beginning, *The Man Without Content*.

The history of art as it enters modernity is, Agamben argues in this text, one in which a strange and unintended emptying of art takes place. His argument returns us to our earlier discussion about the difference between art and aesthetics, where in Hegel we saw a disavowing of art as purely reflective, an *aesthetics* that disunites the subject from the object in a reflective stance over and against the work. Agamben describes the process, beginning in the seventeenth century, by which such an aesthetic moment takes place: one begins to think of and experience art not as embodying the very world and the very self that takes up that world through a comprehensively organic integration (world-work-self), but rather as something that requires one's distanciated, and later disinterested (Shaftesbury and Kant), judgement. Art loses the function of placing the self in its world – of furnishing 'a concrete measure of his existence'[80] – and instead repositions the self in a reflective disembodiment, both of its own being and of the work it beholds. Being becomes a transcendental but groundless ego, while art becomes an aesthetic ideal. Such an ideal might be captured in a term like 'beauty', in which the sensuousness of the artwork, in all its material components, and in the very content that is depicted, gives way to an abstraction and idealisation of something the work is meant to invoke beyond itself. Hence the term 'Beauty' in its capitalised form, a concept that now removes itself from the phenomenal elements and characteristics of the created work that unite and are united with the very selves that engage it – a conceptualisation Agamben calls the 'otherness of a formal principle'.[81] Critical judgement, as aesthetic judgement, we could say *empties* art of its material being, and in doing so empties both artist and beholder of their material embodiment in the world depicted by the art. The artist is thus left to pure creative subjectivity, the beholder to pure aesthetic reflection. Neither now reside, or neither now have placement, in the work, since the work itself is emptied of its content. They reside, rather, in some universal, ideal space altogether removed from the embodying forces inherent within art's phenomenality.

So modern aesthetics is borne out of a negative movement. 'Our appreciation of art begins necessarily with the forgetting of art.'[82] That forgetting, as negation, may be recalled again within the judgement, but it is recalled as the negative, that is, the negated content. Critical judgement 'thinks art as ~~art~~, meaning by this that critical judgement, everywhere and consistently, envelops art in its shadow and thinks art as non-art.'[83] This is the same negative phenomenon that Nancy described above as the vanishing of art in 'Art'. The precise content of *an art* disappears the moment it is conceptualised as 'Art'; it remains latent in 'the Arts', a term which signals diversity, but which still covers over the forgotten sensuous particularity of a singular art with its multiple elements.

Now it is within this context of a negated artistic content that Agamben places the famous clarion call of art's 'end' that Hegel, who informs his discussion throughout, had issued in the Introduction to the *Aesthetics*. For it is not simply a matter of art being eclipsed, made redundant or subsumed by the higher power that is philosophy. It is more that, in a culture where critical reflection has rendered the unity of art wholly fractured, where aesthetics reigns as the principal task of art's contribution to the spirit of the age, where art has been 'transferred into our ideas' and can no longer, *by itself* and on its *own terms*, yield full satisfaction,[84] art must naturally give way to philosophy, which is its only remaining ratification, as now the highest form of the Spirit. Effectively, what Hegel was elucidating, not as prophecy but as induction, was the logical conclusion of an aesthetic move made by his predecessors, and in some cases his contemporaries, whereby judgement, in its disengagement from the Spirit as material, or from the material as Spirit, made ineffectual the capacity of *Geist* to negate itself by, and thereby unite itself with, the material conditions of its day. Only in the *Begriff* was this now possible, a *Begriff* wholly absent from the once potent cultural expression that was art as religion in the *Phenomenology*.

At the beginning of his central chapter, 'The Self-Annihilating Nothing', Agamben references the 'end of art' section in the *Aesthetics* yet again: 'But while on the one hand we give this high position to art, it is on the other hand just as necessary to remember that neither in content nor in form is art the highest or absolute mode of bringing to our minds the true interest of the spirit'.[85] But this is no epitaph. It is not that art is then rendered obsolete; it is rather that art transcends itself: 'Hegel thinks about art in the most elevated manner possible, that is, *from the perspective of its self-transcendence*.'[86] Rather than an end or a death, Agamben claims Hegel implies a 'destiny'. Art's destiny is to go beyond itself, to negate itself. And when art carries itself beyond itself, it 'moves in pure nothingness, suspended in a kind of diaphanous limbo between no-longer-being and not-yet-being'.[87] It is within this state that the titular phrase 'without content' finds its consummation. Art is not decanted of what it contains, or drained as if through some kind of blood-letting. It surpasses itself through its own *kenosis*. But in its self-surpassing it now, historically speaking, awaits its new being, its new beginning. The same, we might say, for the artist.

If the artistic genius is the one whose artistic subjectivity, rent from any specific content in any specific work, is now the *raison d'être* of what it means to 'do' art, the artist is the one who splits most radically and most paradoxically from his defining activity and product. He is emptied of his own content – 'the man without content' – by virtue of that subjectivity which now has no necessary manifestation in the material of his work, for it is his genius that solely marks him, or that 'is immediately identified with his innermost consciousness', and no longer the content of his art. 'The artist is the man without content, who has no other identity than a perpetual emerging out of the nothingness of expression and no other ground than this incomprehensible station on this side of himself.'[88] Historically, this is of his own doing: 'the artistic subject, who has elevated himself like a god over his own creation, now accomplishes his negative work, destroying the very principle of negation: he is a god that destroys itself.' Or, 'Artistic subjectivity without content is now the pure force of negation that everywhere and at all times affirms only itself as the absolute freedom that mirrors itself in pure self-consciousness.'[89] Or again, 'art is the annihilating entity that traverses all its contents without ever being able to attain a positive work, because it cannot identify with any content. And since art has become the pure potentiality of negation, nihilism reigns in its essence.'[90] So that finally, 'The essence of nihilism coincides with the essence of art at the extreme point of its destiny insofar as, in both, being destines itself to man in the form of Nothingness. And as long as nihilism secretly governs the course of Western history, art will not come out of its interminable twilight.'[91]

The young Agamben here is drawing a conclusion about the nature of art in modernity that has surpassed itself through its own critical judgement upon itself, that is, through *aesthetics*, and that therefore awaits its new emergence, its destiny, out of a nihilism that has 'destroyed the very principle of negation'. It is this reading of the history of modern art that, Agamben suggests, Hegel foresaw in announcing the surpassing of art, made impoverished, made empty, by an aesthetics of judgement, or a judgement of aesthetics, inimical to the very heart of negation's generative power.[92] And it is the result of such a history that the young Agamben attempts to resolve in trying to rescue art, and the *poesis* at its core, not only from the lacerating machinery of aesthetic alienation but also from the technical production of modern industrial and post-industrial society that simply re-produces or restates the contentlessness or emptiness of any work that still goes by the name of art.

In his attempt at resolution, or at least at pointing the way out of the destructive nihilism and aestheticism that holds back art from revitalising our place in the world, our placement as the measure of our existence, Agamben solicits others besides Hegel and his 'principle' of negation, as he moves through Nietzsche and his understanding of will, to Hölderlin, Benjamin and, finally, Kafka. But the Hegelian destiny of art, as a destiny that awaits its rebirth, remains very much an undercurrent throughout. And perhaps it comes closest to the surface when, in the penultimate chapter entitled 'The Original Structure of the Work of Art', Agamben begins with a quotation from the late Hölderlin and seizes upon the word 'rhythm': 'Everything is rhythm, the entire destiny of man is one heavenly rhythm, just as every work of art is one rhythm and everything swings from the poetizing lips of the god.'[93] Agamben traces the word 'rhythm' to its Greek roots, and shows how ρυθμός (*rhuthmos*, which stems from ρέω, the verb to flow) was for the ancient Greeks what opposed shapeless, structureless, elemental or inchoate nature. 'ρυθμός is what adds itself to this immutable substratum and, by adding itself to it, composes and shapes it, giving it *structure*.'[94] It also gives it measure, and a temporal measure as much as a formal measure, since what flows flows through time. But rhythm also to a certain extent holds back form and time, or at least holds back specific instances of form and time. For we know that in music rhythm is not something coinstantaneous or reducible to the structure of a piece, nor to the metricality of time signature: 'we perceive rhythm as something that escapes the incessant flight of instances and appears almost as the presence of an atemporal dimension in time.' This transcendence of the temporal instance also plays itself out in a work of art in general:

> In the same way, when we are before a work of art or a landscape bathed in the light of its own presence, we perceive a stop in time, as though we were suddenly thrown into a more original time. There is a stop, an interruption in the incessant flow of instants that, coming from the future, sinks into the past, and this interruption, this stop, is precisely what gives and reveals the particular status, the mode of presence proper to the work of art or the landscape we have before our eyes.[95]

Thus, he says, rhythm both gives and holds back, both propulses and arrests. As the original structure of and for art, then, this primordial kind of rhythm is both a presence or a presencing and an absence or an absenting. We return again to the metre without poetry of Nancy, that is, to an *Aufhebung* in which negation is generatively operative before any and all instantiation. If the original structure of art, long before its alienation in aesthetics, is a certain rhythm that gives and takes away, then this rhythm is a force yet without articulation, yet without meaning, yet, we could say, without content, even as it plays on our senses and places us in a (negative) space that waits to be embodied or measured with precise bearings, contours, forms or signatures of time.

The sense of rhythm here is homologous to Agamben's sense of Voice in the later discussion on negativity and language. It is not merely a between state. It has an original function. For the young Agamben, this original function remains in need of recovery, like the gods of the Eleusinian silence in Hegel's own poem. Poetry, as *poesis*, must guard this silence, as nothingness. But in order to do so, it must return to its kenotic rhythm of surpassing itself to reinstate itself, which it can only do by going through and beyond the alienating effect that inheres in aesthetics. This is to say that if art is to fulfil its Hegelian destiny, art must go through *both* negations modernity has placed before it: the nihilistic negation (Nietzsche's passive nihilism; the will that wants nothing) of an aesthetic judgement that distanciates art from its embodied placement as world-work-self, and leaves no place for the rhythm of negativity to flow and counterflow, rendering art at its 'end' or 'death'; and the generative negation (Nietzsche's affirmative nihilism; the will that wills Nothing) that resuscitates that rhythm and restores us to our original 'site', the original measure of our dwelling, or as the older Agamben will say, to the Voice that establishes an 'essential relation' between language and death.[96] But as with the Voice, the recovery is only possible through the impoverishments of the loss. Only in such passage will we reach a site of beginning anew, the destiny of art awaiting its rebirth.

Agamben's later writings give greater attention to the political dimension of this rhythmic paradox, while still retaining an immense diversity of critical and exegetical focus (philology, medieval science, literature, linguistics, religion, politics, art, etc.). The optimism of youth gives way to a more sobering diagnostics or 'genealogies' of modern political life especially, and of the econo-political equivalents to alienating aestheticism, now much more insidious, in notions like biopower or 'state of exception'. In the latter, for example, an exceptional dictate requisitions the paradigm from which it stands outside in order to become the rule itself. Here we find negativity still in operation: the self-surpassing of one thing to generate a new condition out of its very opposite, so that what stands outside, by standing outside *itself*, makes 'standing outside' the norm. But such negativity remains caught up in an exercise of sovereignty (and not of the kind Bataille rendered), and therefore is far from the negative space or placement that the young Agamben had called us to restore in its 'voiced' silence. The politicisation of negation's creative potency is what, we might say, comes to dominate the *fin de siècle* of the last century, and its new set of conditions in our present era.

SLAVOJ ŽIŽEK

No one has written more voluminously on Hegel and his negation in relation to this new set of political, economic, psychological and cultural conditions than Slavoj Žižek. Like Agamben, Nancy and Kristeva, Žižek's focus extends to many domains, but the political domain, in which ideology is ever operating, is never not present. Nor are two framing figures: Hegel and Lacan. One or the other, and sometimes both simultaneously, inform virtually every page of Žižek's writing – an informing that is also an *in-forming*, a formation of thought always by means of a Hegelian or Lacanian 'apparatus', to use a term Žižek frequently deploys. Such a formation operates both inwardly and outwardly, that is, both within Žižek's sprawling prose and within the Western socio-cultural experience he discusses in that prose. And in this manner, the manner of Hegel's in-itself and for-itself, and the manner of *Begriff*'s formation within the subject and the external world concurrently, Hegel retains a privileged place.

We could thus say that Žižek, as the self-professed Hegelian, is saturated with Hegel. Yet this saturation is also a diffusion. Hegel is both fused into the very fabric of Žižek's thought, and also dif-fused across the very fabric of Hegel's own thought. Like all others we have addressed above, Žižek reappropriates Hegel in order to out-Hegel Hegel, and does so in the name both of surpassing Hegel and of going to the very kernel of Hegel, both of refuting Hegel and of penetrating the 'truth' of Hegel. Such reappropriation is possible only by way of a thoroughly cogenitive 'negation of Hegel'. Žižek tarries with Hegel, but he does so by not staying with him. And this is manifest in his very style: Žižek's writing is itself diffuse, spreading not only across multiple terrains at once, but across lines of argument that easily and quickly divert through one case in point after another, and with recycled ideas or anecdotes that keep re-surfacing in new (and sometimes not so new) contexts. Yet underneath all this diffusion there runs a singular impetus. And however variously this impetus might be labelled, whether Hegelian, Marxian, Freudian or Lacanian – and more likely a blending of all[97] – the impetus takes shape in a consistently particular way. We might label this way more specifically: an 'inversion of perversion', keeping, once again, the full bilateral force of the cogenitive in play.

Žižek is continually inverting our normal way of understanding certain forces at work within our society by stating their 'truth' as counterintuitive. Such statements are rendered with almost formulaic expression, along the following lines: 'so that rather than confirming (the more obvious and expected) X to be the case, it confirms its very opposite, not-X'. A very simple example is offered in the context of capitalism: 'The populist slogan "Save Main Street not Wall Street!" is thus totally misleading, a form of ideology at its purest: it overlooks the fact that what keeps Main Street going under capitalism *is* Wall Street!'[98] But more sophisticated versions also abound. As an example of the socially embodied superego, he writes: 'The point is thus not that the split public/private is not possible, but that it is possible only on condition that the very domain of the public law is "smeared" by an obscene dimension of "private enjoyment".'[99] Or, using literature: 'So – back to Diderot's *Rameau* – the problem with Rameau's nephew is not that his perverse negation of his dignified uncle's "noble consciousness" is too radical and destructive, but that, in its very excess, it is *not radical enough*'.[100] Or in the context of religion, he writes that, for the Jews, 'the Law itself unplugs us from daily rules/regulations – in and through the "unplugging", we do not engage in the orgies that suspend the Law, we encounter *the Law itself* as the most radical transgression.'[101] At one point, he even shows how this formula of inversion would work upon the equally formulaic 'logic of falsification' put forward by Karl Popper (i.e. 'If conclusion *C* is derivable from system *S*, and if *C* is falsifiable, then *S* also is falsifiable'). Žižek writes: 'far from falsifying the rule, the exception one has to search for *confirms* it.'[102] This formula – the inversion of the expected, or in the last case, the inversion of an inversion – is deployed repeatedly by Žižek to suggest a fundamental twist operating within or upon our world of expectations. (Hence his employment of jokes.) And this twist, which twists around virtually all our experiences, is at root Hegelian. So in the last example, Žižek immediately deploys another of his ubiquitous prose devices, 'in Hegelese…', to show the Hegelian translation of the inverted phenomenon: 'In Hegelese, such exceptions are necessary if rules are to become "for-themselves", not merely a natural "in-itself" – that is, if they are to be "noted", perceived "as such".'[103] But this is not just 'Hegelese'; the inversion is not possible without the force of Hegel's negation operating underneath the phenomenon. It is not simply that we can translate the phenomenon into Hegel's terms; it is more that Hegelian forces of negation are forever at work to make both the inverted phenomenon and its translation possible. And it is more, to deploy Žižek's own *modus tollens*: the translation is in fact not a conversion of a set of circumstances into Hegelian principles; the translation itself, *qua* translation, *is* a Hegelian principle (the inverted principle of negation, of the power to be taken up in another mode of expression against one's very own nature, as in the 'cunning of reason') manifested before our very eyes.

For Žižek, these inversions lead to, or become synonymous with, a perversion. What is twisted, counteracted, turned into its opposite, is also, by that very reason, distorted, made deviant, or corrupted from its original or customary intention. Perhaps the most obvious example here is Žižek's reading of Christianity. In *The Puppet and the Dwarf* (2003), which is subtitled 'The

Perverse Core of Christianity', he argues for an immanentist or materialist 'core' to Christian belief, one that is predicated, not on a union with a transcendent Other beyond this world, but, with implied Hegelian inversion, on a dis-union with or separation from that Other:

> We are one with God only when God is no longer one with Himself, but abandons Himself, 'internalizes' the radical distance which separates us from Him. Our radical experience of separation from God is the very feature which unites us with Him – not in the usual mystical sense that only through such an experience do we open ourselves to the radical Otherness of God, but in a sense similar to the one in which Kant claims that humiliation and pain are the only transcendental feelings: it is preposterous to think that I can identify myself with the divine bliss – only when I experience the infinite pain of separation from God do I share an experience with God Himself (Christ on the Cross).[104]

We can see here, in anticipation of Altizer,[105] that the perversion, a deviation from the orthodox understanding of the Cross, is also a certain inversion: the God of abandonment is also the closest, the most intimate God we can experience. But by this very fact, the inversion inverts the perversion itself, so that the experience is no longer 'perverse' in the corrupting sense, but is truer to the very 'essence' of what Christian belief has always tried to reach.

This kind of Hegelian inversion of perversion, or perversion of inversion, which plays itself out across countless contexts and situations, religious, political, psychoanalytic and otherwise, is what we might further call, using Žižek's own term, '*embodied negation*'. Negation is not only embodied in the oft-repeated twist of his inverted, counterintuitive conclusions, with their twisted results, but also embodied in the numerous examples, observations, illustrations, works of art, jokes and anecdotes he reverts to at every turn. Žižek's texts are, we might say, the very embodiment of a negation perpetually and restlessly at work, negation manifesting itself in the material substance of one event, one example, one phenomenon after another, just as *Geist, Vernunft, Begriff* are made manifest in the material substance of this world ('Spirit is a bone', to use the most extreme Hegelian image recycled repeatedly by Žižek). The diffusion of Žižek's prose is precisely the diffusion of negation that permeates the writer's thought, just as the diffusion of negation in Žižek's thought is precisely the diffusion that characterises the form of his prose.[106]

This is not to say that Hegel and his negation are never treated in any specific or isolated manner. In, for example, *Tarrying with the Negative: Kant, Hegel, and the Critique of Ideology* (1993), an entire chapter is devoted to Hegel's 'logic of essence'. But even here, as Žižek maps the middle section of *Science of Logic*, the doctrine of essence, onto a theory of ideology, Hegel is superimposed on various other thinkers (Lacan, inevitably), on various experiences (love, typically), on history (political history, especially) and, as always, on films (such as Chaplin's *The Great Dictator*, unsurprisingly), so that Hegel, tarried with as he might be, nevertheless continually gives way to, or allows himself to be overlaid by, other embodiments, other more concrete expressions. In this sense, one might argue that Hegel is no more or no less present here in the chapter that goes by his name than in any other chapter that makes up *Tarrying with the Negative*. (He is, of course, more statistically present here than in other chapters, if we count the number of times the name 'Hegel' appears. But this misses the point. It is Hegel 'made negative' that is most significant, the Hegel that is subtended, sublated, even sublimated, in the cultural phenomena, the political realities, the notional theories that run their course throughout Žižek's text and the world he describes to us.)

In *The Ticklish Subject: The Absent Centre of Political Ontology* (1999), Hegel once again receives specific focus in the larger context of modern political experience and philosophical thought. Here Hegelian negation (even as the 'negation of negation') is used to expose the void at the core of political subjectivity, in which the Self is hollowed out by its own negating activity. Even the reversal of negation operating upon itself, in negating its own negativity, does not restore the Self to a more whole, more unified, more self-coherent existence. 'Negation of negation,' Žižek says, 'presupposes no magic reversal; it simply signals the unavoidable displacement or thwartedness of the subject's teleological activity... negation of negation *is* the very logical matrix of the necessary failure of the subject's project.'[107] The point of the entire text is not to celebrate this failure as a failure of the Cartesian *cogito*, and thus to dispense with that legacy altogether, but in fact to show, in typical Žižekian manner, that the very 'essence' of this *cogito*, when fulfilling its ownmost mandate, is its 'forgotten obverse', an inversion of its own perversion: the disrupted, dirempted Self as the necessary 'foundation' of all subjectivity, political and otherwise, a Self and *cogito* we estrange only at our peril.

In unveiling this disruptive power of the negativity functioning at the centre of the *cogito* itself, Žižek unleashes the excessive nature of negation that goes beyond the dialectical structure in its standard rendering. Self and Other are not simply mediated as Self-for-Other. Other comes to populate the very essence of Self through a negation of negation that extends beyond the mere role of mediation. That is, two opposing sides are no longer simply triangulated, but rather are opened up to each other's internal multiplicity. In one section ('3, 4, 5'),

Žižek offers his own reading of the 'fourth term' Kristeva had elaborated earlier for us. Referring to the same closing passage of the *Science of Logic*, where a triplicity carries the possibility for a quadruplicity, he posits the idea of an *inter*subjectivity missing from the normal triadic function of the dialectic.[108] He finds this possibility in the very structure of the *Greater Logic* itself, which, we recall, is built around the bi-fold division of objective logic (Volume One: doctrines of being and essence) and subjective logic (Volume Two: subjectivity, objectivity, and idea). But should there not be a third element to the structure, Žižek asks, a sublation of objective and subjective logic, just as in the second volume subjectivity and objectivity give way to idea? Is there not something that stands waiting to be reconciled between these two, something that follows 'the overall articulation of the dialectical process in which subjectivity comes second and stands for the moment of split, negativity, loss'?[109] This subjective logic then is to give way to what he calls an intersubjective logic, whereby the movement between Substance and Subject always invokes a fourth phase. So (working backward from the *Logic*'s structure), an idea (1) becomes externalised in nature (2), then returns to itself in the finite subject opposed to nature (3), and then is reconciled in 'an intersubjective *Sittlichkeit* as a man's "second nature"' (4).[110] This intersubjective logic – 'the I that is a We and the We that is an I', as Beiser reduces it – might even give way to yet a fifth movement, an absolute logic.[111] However diffusely elaborated this argument might be presented to us (and in some ways Hegel does address it in his reworking of the *Logic* into tripartite volumes, whose third volume, *Philosophy of Mind*, incorporates that very intersubjective *Sittlichkeit* in 'Mind Objective', before it becomes 'Absolute Mind'), the point Žižek wants us to take away is that negation continues to be excessive even to itself. What is supposed to 'contain' it – a tidy, mechanical triadic dialectic – is itself internally combusted by negation's own machinations.

The culmination of Žižek's interaction with and embodying of Hegel, in an excessive manner that goes beyond Hegel, is the large volume of over a thousand pages, *Less Than Nothing: Hegel and the Shadow of Dialectical Materialism* (2012). Here, as with the others above, the intention, now explicitly stated, is to read Hegel by out-negating Hegel, but thereby staying most true to Hegel, and bringing Hegel back as a properly acknowledged force within our present-day conditions, even if the force itself has to be re-constituted by its own negativity. At the centre of Hegel is Hegel's own split, just as at the centre of all subjectivity is our riven core. And this insight is Hegel's uniqueness, even among all German Idealists. It constitutes our primary 'ontological failure': 'what appears to us as our inability to know the thing indicates a crack in the thing itself, so that our very failure to reach the full truth is the indicator of the truth.' This is as true of Hegel himself as of any other phenomenal 'thing' we might observe; and it is why this negative insight remains the premise of the entire thousand pages. It is also why 'Hegel has lost none of his power today', even if, here in the text, pages upon pages can still go by without ever a mention of Hegel.[112]

Less Than Nothing is thus not merely Žižek's most sustained treatment of Hegel; it is also his most sustained performance of embodied negation. Everywhere throughout the text, negation is at work in the cogenitive form of 'the negation of Hegel'. One can even see this in the structure of the text itself (very aware of Hegelian textual structures as Žižek has proven to be). The entire book is organised around a sexual metaphor, or a sexual witticism: 'It was said (in the old days...) that the second and third most pleasurable things in the world were the drink before and the cigarette after.'[113] Part I of the book is entitled 'The Drink Before', Part IV 'The Cigarette After'. The two parts in the middle – the unstated act of copulation – are respectively 'The Thing Itself: Hegel' and 'The Thing Itself: Lacan', as if these two partners were in an embrace with each other, as they enact the 'thing itself', the sexual union. Parts I and IV have three chapters each (with many subsections). But the middle parts, II and III, have only one chapter each, yet are followed in both cases by three 'Interludes', the third of which involves either sex (Part II, Interlude 3) or sexual difference/sexuation (Part III, Interlude 6). Deliberately, then, the enactment of the 'thing itself' is overwhelmed by interludes, as if through a repeated *coitus interruptus*, just as the term 'sexual intercourse' is withdrawn in the witticism itself. That is, as in the psychosomatic libidinal economy, the greatest moment of consummation is obscured by the surrounding eroticism of deferment, suspension, intermission or afterglow, to suggest that the climax of erotic ecstasy is not the moment in which one engages in the sexual act itself, but the ecstasy of the ecstasy, the standing outside the copulative moment, what Žižek calls in the second Interlude 'the ex-timate core'.[114] Such extimacy (countering intimacy) is in fact *inherent excess*, the excess of Hegel that is Lacan, and of Lacan that is Hegel. But it is also the inherent excess of the Hegelian system itself, as triadic formulations give way to a (at least) *quadral* system of dynamics. So there are *four* parts to the book; the two middle parts have *four* chapters each (the latter three of which are interludes, supposed interruptions of the 'thing itself'); and the final part, though constituted by three chapters like the first, deals explicitly with (in the first two chapters) 'foursomes' or with (in the last chapter) 'quantum' phenomena. And yet all of this 'structure' appears lost in the actual text itself, since in content neither chapter nor interlude seem suitably differentiated from one another, and since Hegel and Lacan (*the things themselves*) are forever diffused, displaced, sublated or sublimated by a continuous discursive flow that at any one point may be 'quantized' by another figure or another theory, by constant reversion to an example, to an 'apropos', but that ultimately exceeds any one discrete expression or proposition or postulate. This is the pure embodiment of negation, then: the thing itself (Hegel, Lacan, their intercourse, etc.) is 'itself' only by means of a negated embodiment, which is its true embodiment (in another).

To avoid the risk of succumbing to Žižek's textual excess, let us take but one example from his pages. In a section of Interlude 1, Žižek once again unpicks the standard reading of Hegel's 'negation of negation'. Many commentators, still to this day, understand this self-reflexivity of negation as part of a forward-moving advance to a higher state, by which negation, in working upon itself, un-posits the negativity that the first negation brought, and in turn posits negativity's negation now as *something* (in contradistinction to nothing). This something, as Something – the higher state of the triadic movement (traditionally, the synthesis) – arises through negation working solely in a *mediating* capacity. Negation mediates between something and nothing, but, by virtue of its role as agency, it effects a Something that, even if it owes its existence to negation, has left its nothing behind. But this, argues Žižek, misunderstands the internal dynamic of self-relation in the 'negation of negation'. Yes, negation is always and only ever *in relation*; it requires something on which to act. But this does not mean, as is so often assumed (e.g. by Sartre), that negation can only exist posterior to something positive, to positivity that first receives its negating powers. True self-relation means that the *something* on which negation first acts is none other than itself as negation (before it is even nothing). This is to say that the something *qua* negation, which is subsequently negated, already has its somethingness disturbed or disrupted by the negating power that, rather than being received afterwards, in fact has always been operating within it. Something is always and already its own negation. Hence the inherent fourth term: there is something (1) that already and necessarily carries its own negation (2); that something, *qua* negation, is then negated (3); that 'negation of negation' brings a 'higher' form (4) that is born with the original negation still fundamentally at its core. We are thus not in a 'higher' state of now purer positivity; we are 'higher' (which in fact is 'lower') in a state (which in fact is no state, but an ongoing dynamic) of *negation*, which we could even now construe as an absolute negation (5).[115]

If this is confusingly abstract or, in defying common-sense reasoning, 'meaningless sophistry', since 'one cannot begin with negation' because negation assumes, by definition, an object to negate, then it is crucial, Žižek tells us, 'to explain what is meant by the self-referential negation *through convincing examples*'.[116] In other words, if this reading of 'negation of negation' is to hold, it must be shown that the original starting point of negation is 'something' that can be exemplified with(in) its own negation – a historical account, a concrete situation, a literary passage and so on. And we are given a full range of such examples in what immediately follows: Marx's critique of bourgeois freedom and equality (which already carried within it its own immanent self-negation of freedom *before* any revolution was imposed upon it); a passage from G.K. Chesterton's *The Man Who Was Thursday* (in which law already has crime resident within it, and vice versa – a variation on Agamben's 'state of exception'), which is supported by a quotation from Heinrich Heine's *History of Religion and Philosophy in Germany*; a quotation from Richard Wagner's draft of his play *Jesus of Nazareth*; a reference to Marx's *The Communist Manifesto* (bearing in mind the interlude in which all this comes is entitled 'Marx as Hegel, Hegel as Marx'); a passing reference to Louis Althusser's *La revue de l'enseignment philosphique*; a more extended reference to Jung, in his opposition to Freud; an ever-risky example of the struggle between Nazis and Jews; and finally a comparison to Zen Buddhism, and its (differing) understanding of the Void.[117] What we see here is a typical embodiment of Hegel and negation in something other: the 'tangible' outworking of the Spirit that is the negation of Hegel, the Spirit that is the 'shadow' of the book's subtitle ('The Shadow of Dialectical Materialism') as it casts itself outside itself in negation of itself and in some other 'material' form. By the end of this section 'Negation of Negation', he will speak finally of 'the properly dialectical paradox of a Nothingness which is *prior* to Somethingness and, even more, of a weird Something which is *less* than nothing'.[118] This 'even more', which becomes 'less than', is nothing other than Negation itself, which is even prior to the anteriority that is Nothingness to Somethingness. Here, 'negativity is not reduced to a self-mediation of the positive Absolute,' but, on the contrary, 'positive reality appears as a result of self-relating negativity'.[119] Yet even here, though he finishes with these more abstract assertions, they come to us mediated through the foregoing examples, and in fact end on yet one more example, now from the domain of ethics: 'the good is a self-negated or self-mediated evil'.[120] In this self-mediating sense (the self-mediating of/as self-negating), Žižek is arguably more *performative* of Hegel than all those we have discussed above. Or at least we might say this, as translated into Hegelese: his pervasive performance of negation has reached the state of being in-and-for-itself in its pervasion.

But what, finally, of the *creative* side to this performative nature? Is this ever made explicit in the text, or just performed? Is art ever an object of concern itself, or does the political and the ideological remain the ruling framework for all this performative shadow-casting? There are occasions when Žižek addresses the topic of art, though they are comparatively few. In one of the more salient examples, he engages with Robert Pippin's reading, in Pippin's *The Persistence of Subjectivity*, of modern abstract art as an extension of Hegel's narrative about aesthetics. Rather than art remaining surpassed by the absolute knowledge of philosophy, Pippin suggests that art re-emerges in a manner consistent with the self-reflexivity of that absolute knowledge, now no longer in a representational mode, which was beholden to a transcendent sphere underwriting the subject–object split, and which kept art to a mediating role, mediating a reality it could never fully capture, but rather in an abstract mode, by which art reflects on its own art-ness, or, as he says more specifically, painting on its own 'painting-ness'.[121] Žižek accepts this reading as a 'radicalising' of abstraction, art's own 'reflexive questioning of the very medium of artistic representation, so that this medium loses its natural transparency', and by this agrees that Hegel's narrative remains operative within modern(ist) art (even if not postmodern art).[122] But he does not directly endorse the creative power of negation in this operation (as Pippin does not). That endorsement perhaps comes later when, in the context of Kant and Deleuze, he offers, in the form of a question, what might serve as the encapsulation of the book's entire 'thesis' on Hegel: 'what if the distance of re-presentation, the distance that renders the Thing inaccessible to us, is inscribed into the heart of the Thing itself, so that the very gap that separates us from the Thing includes us in it?' And immediately at the beginning of the next paragraph he says that the 'exemplary case of such a creative process is *art*'.[123] So Žižek acknowledges here that the very gap we think makes us distinct (and limited) in fact manifests itself as our very own selves, and that this manifestation is more than just a showing, more than mere re-presentation: it is at the very core *creative*. So in the context of art, when we say that the 'Thing' is the reality art is trying to represent, then the gap that it cannot overcome is precisely what characterises the nature of that Thing, so that the gap, as negation, is not only what is re-presented, but more importantly is what allows re-presentation in the first place, starting with a re-presenting of itself.[124] When, however, the 'Thing' is Hegel himself (or Lacan himself), as the section titles indicate, then what separates us from Hegel's thought today is precisely what we are inscribed into *as* Hegel's thought, the thought of the gap as negation, and we are closest to Hegel precisely when we embrace this gap, when we reside in it or live out of it or, as *Glas* showed us, write by means of it.[125]

But it is not any one clear or sustained explication of this creative process that marks Žižek at his most Hegelian in relation to the art of negation. It is, as we have already said, his very embodiment of this process. Žižek is most Hegel-like, or is most like the inverse perversion of Hegel, which, as Žižek everywhere implies, is the 'true' Hegel, that is, the self-consistently self-negating

Hegel, when he embodies the negation of Hegel through the embodiment of exemplary works of art. Every literary, cinematic, operatic and visually artistic illustration or reference is not merely an attestation of Hegelian forces at work, but the creative forces themselves, embodied in the text (of the art and of Žižek alike). And this is exactly why 'art' for Žižek retains none of the traditional boundaries: popular culture can be as much the embodiment of these forces as supposed 'high art'. Because art is no longer reflecting some transcendental ideal (Beauty, Goodness, the image of God, etc.), but reflecting instead ourselves as self-reflexive doers and thinkers, and thereby reflecting the gaps that are these very selves, as reflexive (the self-relation of our inherent internal negation),[126] our self-reflexivity comes in all forms, from the Hollywood spectacle and saccharine TV chat shows, to our restaging of Greek dramas and Wagnerian operas. Indeed, this creativity need not stop at 'art' per se; it extends to all social, political, economic, intellectual and psychoanalytical activity we produce from within our culture, whether 'high' or 'low'. (And that the 'low' can exemplify the 'high' confirms the point: the perversion of inversion is constantly at play.) We might say Žižek is quintessentially Hegelian in this regard, as his texts showcase to us the spirit of our world, in all its stream-like variation, made concrete in the discrete moments of an isolated case. Yet the rub lies here: the 'spirit' he makes manifest is Hegel himself, but of course, and at the very same time, not Hegel *himself* – Hegel as always somebody other, Hegel as non-Hegel, Hegel even as *less than nothing*, which, in the profundity and perversion of the paradox, is always *more* Hegel than we bargained for.

CATHERINE MALABOU

We cannot conclude our contemporary tour of re-Hegelianisation without some consideration of Catherine Malabou's work, and particularly her text *The Future of Hegel: Plasticity, Temporality and Dialectic* (2005; orig. 1996). For not only does this work return us to Derrida, who supervised the doctoral thesis that eventually became this her first major publication, but it also, in its very circling back, takes us forward beyond Derrida, and in doing so, forward to Hegel beyond, with a temporality that is purposively inverted.

Malabou's Hegel, like that of so many of her Continental contemporaries, is a complex and fully reconsidered Hegel; but negation is not always the clear and present danger it is in, say, Žižek. Moreover, her shift towards science, and neuroscience in particular, might suggest she moves away from any art inherent in that negation. But the Hegelian concept central in all her work, as it first flowered forth from *The Future of Hegel*, that of 'plasticity', very much owes its lifeblood to art and aesthetics, and thus if we concentrate on this initial first text we gain a crucial sense of how her subsequent work becomes an adaptation of Hegel's art of negation uniquely defined, a philosophising of negation 'in one's own idiom', to borrow directly, and ironically, from Hegel.

(And so 'first texts' become hugely significant in this coterie of contemporary *Aufhebungs* of Hegel: Kristeva, Nancy, Agamben and Malabou each begin their writing career with Hegel and negation; and if Žižek's bibliography cannot evidence a singular departure point in Hegel, this should not surprise us, because Hegel, as in all his work, hovers over the waters of his early Slovenian texts in both stated and unstated form.[127] We can therefore say that, even here with the emergence of text, and *especially here*, Hegel's negation becomes a generative force, the start of reading Hegel anew in one's own idiom, the start of Hegel's own finish.)

The very beginning of *The Future of Hegel* sets out a problematic: is Hegel himself, or is Hegel's philosophy in any form of a Hegelianism, 'a thing of the past'? Right away, we begin in the spirit of art and the *Aesthetics*. The first moment in the beginning of a triadic movement – art < religion < philosophy – becomes the all-consuming moment for the question of Hegel. What has a future in Hegel? Is Hegelian philosophy a thing already surpassed, and therefore precluding a future? The question of time and temporality will figure as part of Malabou's own triadic approach, as she moves to address this opening problematic in the present. So too will the question of the surpassing itself, which must involve a dialectics of time and tense (the past, present and future of Hegel, cogenitively). But at the beginning of this movement (as announced in the subtitle) stands *plasticity*. And this concept, as concept, has its direct correlative in art.

The term 'plasticity', with its cognate forms (plastic, plastics, etc.), appears frequently enough throughout Hegel's works (as *Plastizität*). But its most relevant occurrence is in relation to the arts, as the *plastic arts* that gain attention in the context of Greek art in the *Aesthetics*. What Hegel has in mind (following Schiller) in the section he entitled 'The Classical Form of Art' is, as its second chapter tells us, 'the process of shaping'. And materially, sculpture is the paramount expression of this process, whereby the material lends itself to being shaped and moulded.[128] But this lending is only just that: what the material gives of itself, or gives up of itself, is only its plasticity, and not its essential nature. As plastic, the material retains its own essential substance as physically constituted – the clay or the marble is not transformed into or mixed with some other substance, but keeps its original material constituency – and yet it yields that essence to a being moulded otherwise than its indigenous shape. This simultaneous reality of obduracy and pliancy, of retention and cession, is the predominant characteristic of plasticity.

We have seen time and again how the simultaneous retention/cession of opposites marks Hegelian thought. So it is no surprise that Hegel would seize upon the Greek classical form as the beginning of an important move towards a higher manifestation of Spirit. For it is in the dynamism of retention/cession at play between the material and the immaterial that both a truer individuality/subjectivity and a truer universality/Spirit are allowed to emerge. Thus in the *Aesthetics*, Hegel can claim that 'the perfect plasticity of gods and men was pre-eminently at home in Greece.'[129] There the subject and Spirit find a material substantiation not before seen, and Greek art becomes a paragon of its process and its expression, even if the material does not yet liberate the Spirit to its truest independence. At the end of the *Aesthetics*' first volume, Hegel summarises this Greek realisation: 'In the plastic figures of classical art the subjective inner element is so related to the external one that this external is the very own shape of the inner itself and is not yet released therefrom into independence.'[130] So to be plastic means, for Hegel, to possess a double and contradictory nature that co-penetrates between one's inner and outer being.

For Malabou, the term 'plasticity' is therefore malleable across Hegel. For it does not simply begin and end with aesthetic features of an era gone by, but comes to mark the entire philosophy of Hegel at its innermost conception and operation. In Malabou's conception, its double nature is not only that it retains material substance while it allows that substance to be remoulded, but more that it has also 'the power to bestow form, the power to mould' in its own right. To be plastic, to possess plasticity, is *'being at once capable of receiving and of giving form'*.[131] The double nature therefore works across (at least) two levels: the substance/form dichotomy, and the material/spirit dichotomy, levels which of course are inherently related. For though, on a material level, the plastic substance allows itself to be re-formed by external hands, on a formal level its substance gives form to a spiritual reality. It thus receives form from outside itself, but gives form to that which lies beyond the material. Malabou thematises this doubleness in the term *substance-subject*: 'substance withdraws from itself in order to enter into the particularity of its content. Through the movement of self-negation substance will posit itself as subject.'[132] In this sense Malabou, following Hegel, can speak of a 'plastic subject' or of 'plastic individuality'. The subject forms, or begets, its own substantive content.

We have already seen this plasticity in operation throughout our foregoing discussions, and in various manifestations. For one, it operates in a similar manner to Schiller's *Spieltrieb*: the combination of life and form that yields a 'living shape' through a dialectical process of sensuous and formal impulses. This shape, as Beauty, is, we might say, the plasticity come to life, both materially and conceptually. When Malabou later describes Hegel's anthropological understanding, it in fact accords identically with Schiller here: 'Human characteristics are not a given: they emerge as the result of a process of formation of which art is the paradigm.'[133] In another manifestation, plasticity can be seen as an isomorph of our own repeated term – cogenitivity. To be both subject and object at the same time, to hold both subjective and predicative status bilaterally, and yet to go further and generate the one from the other, is precisely what Malabou's plasticity opens up in all its directions and extensions: 'Elevated into its speculative truth, the relation between subject and predicates is characterized by "plasticity"'; 'Self-determination is the movement through which substance affirms itself as at once *subject* and *predicate* of itself… the very plasticity of substance itself, its capacity to receive and to give form to its own content.'[134] Such is the *plasticity of Hegel*. Hegel forms, performs, out-performs, informs, deforms himself by his own plasticity. And such plasticity therefore marks the dialectic of Hegel, the *Aufhebung* of Hegel, the negation of Hegel.

The Hegelian dialectical movement is therefore plasticity at work across all levels, generating formation across the technical (*techne*), the cultural (*Bildung*) and the authorial (Hegel), even across all at once – the *formation of Hegel*. Malabou writes:

> The dialectical process is "plastic" because, as it unfolds, it makes links between the opposing moment of total immobility (the "fixed") and vacuity ("dissolution"), and then links both in the vitality of the whole, a whole which, reconciling these two extremes, is itself the union of *resistance* (*Widerstand*) and fluidity (*Flüssigkeit*). The process of plasticity is dialectical because the operations which constitute it, the seizure of form and annihilation of all form, emergence and explosion, are contradictory.[135]

So Malabou sees plasticity as a contradictory movement, but a movement *through time*, so that when the plasticity becomes the cogenitive plasticity of Hegel, then it necessarily raises the question of Hegel's time and Hegel's future. In their triangulation, plasticity, dialectic and temporality become the contradictory future of Hegel as a thing of the past – not a future perfect, but the present *formation* of the future of Hegel that is beyond all tense.

Now where does negation enter into this future? Plasticity, we have seen, already carries within its conceptuality a negative impetus: in its doubling, it opposes its own internal nature (and temporality). But that opposition, we have also seen, instigates its own formation. Plasticity, then, is both a formation and a formulation of the negative; it both takes and receives *negation*, while at the same time it gives a sense to negativity. At various points throughout her text, Malabou will try to give both this formation and this formulation a context. In, for example, her first part on anthropology (an exegesis of Hegel's first section in the *Encyclopaedia*'s *Philosophy of Mind*) she will read Hegel reading Aristotle on the concept of *Nous*, not merely as mind but as a 'way of being' that involves a 'differentiated unity', an auto-differentiation that programmes negativity (as self-division) into the very act of being oneself.[136] Or in the second section on Hegel's God, and the death of this God, the kenotic will feature centrally in marking out a divine plasticity whereby the God who posits Himself in exteriority (substance-subject) becomes alien to Himself, but who thereby can now be *seen* in that alienation, both aesthetically (depictions of the incarnated Christ in Romantic Art) and self-reflectively (the God who knows Himself as passed over into the world as *Geist*, the coming together of a God and humanity who 'see themselves pass by').[137] Both these examples are rich veins for plastic negativity.

But it is in the third section, on Hegel's philosopher, that we get the most cogent sense of negation as a creative force. And here the discussion returns not merely to reading Hegel's thought in-itself, but to reading Hegel's thought for-itself, towards an in-and-for-itself. For in the last part of this third section, Malabou turns to the act of reading Hegel, an act that must now bring all the foregoing questions of plasticity into force.

In moving from formation to formulation, the philosopher who thinks, who enacts thinking, must shape not only philosophy's content but also its form, and do so towards Absolute Knowledge. It is the philosopher who supremely embodies 'thought which begets itself as thought'. But since that thought can only perform its supreme function under the full formation of subjectivity, a formation only Absolute Knowledge can bring, the philosopher is caught in a hermeneutical circle, or a 'speculative hermeneutics' that 'must consist of a mutual gift of form' – the subjective philosopher must become an absolute subjectivity, but only through absolute subjectivity can the subjective philosopher be formed.[138] The formation here must therefore be of the plastic kind if it is going to work itself around or break through into this vicious circle.

Malabou translates this circularity into the problem of the philosophical proposition. In traditional exposition, the proposition works in a linear fashion from subject to predicate via the copula. For Hegel (in the Preface of the *Phenomenology*, §23 and §59–62), the subject in the traditional proposition becomes the philosophical subject, that is, the philosopher as subject, who takes over the subject as the knowing 'I', and allows predication to follow subsequent to its grounded subjectivity as knowing. But such linear transition meets with a hermeneutical and grammatical resistance. Malabou refers to Hegel's example of 'God is being': the copulative verb here cannot maintain a unidirectional transition, since both 'being' and 'God' vie for the status of subject, and predication is subsumed by, or falls away to, the subject as knowing Subject. Although analytical philosophers and language theorists have put forward their own solutions to such propositional difficulties, Hegel's solution is *speculative*: rather than harmonising subject and predicate, grammatically or conceptually, the speculative proposition destroys any distinction, and produces a selfsame identity, so that the predication process becomes cogenitive – proposition becomes preposition. Or, as Hegel describes it, the negative belongs to the propositional content *propositionally*.[139]

If the proposition becomes Hegel himself, or Hegel's philosophy, Malabou then asks, how do we read or re-read this proposition? In the philosophy *of* Hegel, how do we read or re-read this preposition? What does it mean to enact a reading of a subject (Hegel) and its predicate (Hegelian philosophy) when the two are so totally absorbed into each other that one does not know how to properly take up the subject? How do we, as now the knowing 'I', or as the reading 'I', take up the subject that is Hegel, and predicate upon it? How do we advance without going around in circles, without returning to Hegel as a subject already formed by his own content? How do we properly negate the proposition that is Hegel, and Hegel's philosophy? How do we *return anew* to Hegel?

Plastically, we must read the proposition of Hegel bi-directionally, or in a circuit that allows passage both ways: giving and taking, going forwards and backwards, presenting the future and the past, the subject and the object, as if in a full identity without reserve – and all taking place at the same time. If the Hegel we encounter is at first unreadable, because the unity of his concepts come about speculatively, dialectically, plastically, which destroys the form of the proposition we hope to read and interpret as Hegel, we must reposition ourselves and return back along the propositional process, returning anew to the subject at the origin. But at that very origin is not the substantial subject of the knowing 'I', the Subject (philosophical or otherwise). Rather, at that origin is a divided subject, who is therefore no real subject at all. As Malabou writes: 'But at the place of return, the reader finds *nothing*. As the origin was never there the first time, the reader cannot discover any substantial presence or substratum waiting to be identified. The only thing that can be followed is the sheer movement of retrocession itself, a return that moves on, finding at its own back the drive to advance.'[140] That drive is in fact the force of negation, so that what really stands at the origin is not nothing per se, but negation, which does not 'stand' there, but *operates* there, acts itself out there, along the circular path of its own very nature.

This is a very circular affair, then, but as much as the movement traces out a hermeneutical circle, in which beginning and ending become confused, conjoined, conflated, it is also an artificer's circle, in which something awaits to be created. So Malabou can continue: 'Turning back to the original point at which all the forms had been presented, the reader is at the same time projected in advance: required to give form. Plunged into the void of the proposition, the reader is brought to formulate new propositions in return.'[141] This giving form, which takes place in a temporal eradication of tense (past, present, future), allows us to return anew to the subject, the content, the concept, Hegel himself, which all now stand to 'be understood differently': 'Because this understanding [of the proposition] was not derived from itself (a first reading has never happened), the reader must have produced it.' And we go on producing understandings, not passively, as if we only *receive* something from the proposition (its objective claims), but plastically, as we form any claim we might venture by means of the pliability that characterises the cogenitivity of the subjective/objective.

If each of the living thinkers above has understood and, in their writing, embodied the returning anew of Hegel, it is Malabou who most thoroughly articulates its projection. And if each of the thinkers above pays homage to the negative creation that this projection entails, by working out (and within) creation's own negativity, it is Malabou who most implicitly (but by the very plastic nature of this implicitness, which implies, even expects, explicit formation) opens up the space for a new art *form*. The formation of this form comes, ironically, with 'art' least stated. But resident within the very plasticity that makes a thing of the past alive again for the future, art, as the art of negation, gains a new freedom.

Derrida writes in his Preface to *The Future of Hegel*: 'To invent, and most particularly understanding invention as an event, means here to rediscover what was there without being there, both in language and philosophy'.[142] To discover in Malabou the art that is there by not being there is one of the great achievements of her book. But of course the 'thesis' is meant to apply first to Hegel himself: discovering anew the Hegel who is there by not being there. And we can extend this now to Derrida himself, who, by informing, we can assume, every page of the original thesis, is there by not being there. Of course, he is explicitly there in the Preface. But what does Derrida's Preface add to the book, and indeed to Hegel, that either Malabou or Hegel has not already said? Derrida finally gives us another writing 'on Hegel', yet that writing is already surpassed by his pupil, and therefore is less a writing on Hegel than a writing on how one writes (of) Hegel. What it adds is only what is not there – much like the central margins of *Glas*.

Living Hegel involves this strange absenting. The art of reading Hegel now is to enter into his absence wholeheartedly, to live out of his own negation by living out his negation, which becomes our own absence, as philosophical authors, as speculative readers. 'Hegel may well be the first philosopher to think that the author is not "a fixed and solid subject" but an instance of writing, conceived in the joint play and speaking of two subjects of enunciation: a speculative reader and the one who wrote because he was, first and above all, a speculative reader.'[143] How precisely we enact this living, and enact it responsibly, is the task that looms upon our future.

PART III

FURTHERING HEGEL

CHAPTER 7

The Ought of Negation

If we are to think about a return of Hegel in today's world, upon the futuring of Hegel that Malabou has laid out for us, then, as we have said repeatedly, and along with Malabou, we must think about this 'return' as a 'beyond'. Despite a logical insistence that returning is mutually exclusive with going beyond, we must understand, as our contemporary chorus of voices have understood, that they are one and the same. To return to Hegel today is to go beyond Hegel. But to fulfil a thing by surpassing it, to embrace a thing by negating it, is *precisely Hegel*. That is, we best serve Hegel, we best work Hegel, when we out-manoeuvre and out-wit Hegel *in his own name*. Yet did not Hegel announce this very phenomenon himself when, in the Preface to the *Elements of the Philosophy of Right*, he gave us that all-too-famous image of the bird whose glaring eyes come alive only after the sun has fallen: 'the owl of Minerva begins its flight only with the onset of dusk'?[1]

German Idealism might have long since had its day. And a Hegelian 'ideal' of that idealism, that a certain absolutisation of a progressive march of history finds itself manifest in the Western culminations of the advanced modern Spirit, is also a hope now left far behind. So why does Hegel return to us now? And why today, of all times, when progressive marches of history have led, and continue to lead, to such catastrophe, to such regressive attempts at how the world ought to be, and how we ought to think the world to be? Perhaps we can say it is precisely because our progressions have become regressions that Hegel returns, but now a Hegel who, we have seen, is continually outstripped by his own internal mechanism. For Hegel invoked Minerva's owl in a paragraph that began this way: 'A further word on the subject of *issuing instructions* on how the world ought to be: philosophy, at any rate, always comes too late to perform this function.'[2] If philosophy is always a step behind the ethical imperative, if the ethical life is already actualised before philosophy can set its own agenda upon it – and this from a book whose idealism will go on to set just such an agenda – then philosophy (as it is formed by Hegel, or as it is informed by Hegel) will always have to *return back* to find itself. But in doing so it will in fact undo itself, go beyond itself as a doctrine of positive contrast to what has gone before. Instead its coloration will be 'grey in grey', as Hegel says,[3] neither white nor black, and not even a grey mixture of both, as in a convenient act of sublation between two extremes, but a grey in a grey, a sublation of a sublation, a 'negation of negation', one that effects a disappearance, but at the same time brings out the colour all the more. It is there by not being there. Hegel returns, even at the point of ethical uncertainty, or even at this point of ethical actuality that carries so much uncertainty, because Hegel is our first modern philosopher who understood the absolute need and the absolute desire to negate himself.[4]

The return of Hegel today in certain circles of philosophy and theory, we have now seen, is 'predicated' on this negation. But since we now know that, in Hegel's words, 'philosophy forms a circle', the return of negation today is predicated on Hegel in a manner that sends the predication back to its origin, which is also its future.[5] Hegel becomes both predicator and predicated, both subject and object, in a returning anew to the idea of a negation beyond what has been called the postmodern turn. What bearings do we give this new territory beyond the postmodern, if Hegelian negation is so central to its concerns? Why negation's insistence, even by those who do not claim, by choice or by perplexity, to be among the Hegelian esoteric? Is it simply that, as Derrida suggested, a neo-Marxism continues to exist in the thought of Western intellectuals, as neoliberalism and its capitalist regime stumbles and wavers? And so if we return anew to Marx, we also, and ineluctably, return anew to an inverted Hegel? Certainly the Continent continues to wrestle with Marx, as Jacob did with the angel at the ford of Jabbok, whether in real or spectral terms. But it is more than Hegel by way of Marx. If postmodernism was the political gesture to limit, in the name of theory, the hegemony of powers that resulted from modernity's aggrandisement of Subject and Object, and the eventual, and sometimes extreme, antagonisms that ensued between them, then beyond postmodernism we find the attempt to re-address those antagonisms, not by way of concord and reconciliation, but by way of *speculation*: the negation of either side that is also the self-begetting of either side – the self-begetting of one's own negation. Both thought that begets itself (as negation), and negation that begets itself (as thought). This cogenitive self-begetting is the very essence of the art of negation. But in its begetting it also raises the spectre of an ethics. For if, in the grey on grey, which is no better materialised for us in the late work of Mark Rothko (the ultimate abstraction of art that Robert Pippin suggests could reach the self-reflexivity of absolute knowledge), philosophy comes too late to issue instructions on how the world ought to be, as postmodern theory came too late in the twentieth century, then what comes after theory, and what comes after aesthetics (and, indeed, after the theory of aesthetics) must address this problem, the problem of the 'ideal' now in its most pejorative sense – that which takes us beyond the actual, beyond the here and the now, that which is wished for, but never seems possible to attain.

We have seen, beginning with Kristeva, how the creation of negation raises an ethical insinuation, and if taken forward an ethical insistence: shattering, dissolving, vanishing, annihilating, surpassing – all these imply not only a power, but, necessarily, a violence. Negation, as the imposition of force, is never outside questions of authority, and therefore never outside questions of responsibility. And we can see in the trajectory of the contemporary thinkers of the previous chapter a move towards these issues, as questions of art lead, inevitably, to questions of community (Nancy's community of 'being singular plural', e.g.), of politics (Agamben's permanent state of exception, e.g.), of religion (Žižek's perverse Christianity, e.g.), and of bioethics (Malabou's focus on neuroscience, e.g.).[6] If Hegel is right, and philosophy indeed has come too late to the socio-political realities of late modern extremism, how can Hegel himself possibly be used to rescue (to alter Hamlet's phrase) a philosophy out of joint, much less by means of an art he himself claimed to be out of joint? By returning to Hegel, how can negation furnish the grounds for what 'ought to be' in the world? Especially when, if we are to stay with Hegel, we would have to return to that 'ought' as something already actualised? How can an ought be actualised? The task here is a steep hill. And, as is often the case with Hegel, one is never quite sure whether one is going up the hill or down it. But whatever direction, traversal is possible only if we see negation, not in terms of a power and violence imposed by the security of its own authority, but in terms of an active, creative, generative force whose only security is that it must give itself up to its own powers. Negation must be seen as the activation of effecting, but effecting the existence of a something that is nothing other than its own internal and unconditional negation. The question we now have before us is whether, in the very nature of this unconditionality, which has no conditions imposed upon it but its own self-negating reflexivity, it can effect into existence an 'ought'.

The two customary ports of call for any question of what 'ought to be' in Hegel are of course the *Phenomenology of Spirit* and, more commonly, the *Philosophy of Right*.[7] In the *Phenomenology*, the section on Spirit, which follows the section on Reason and precedes the sections on Religion and Absolute Knowing, reveals *Geist* first as the ethical order of culture, before in its self-certainty it becomes morality. Here, the moral view of the world develops under the shadow of a Kantian 'as if': though a duty not found in Nature itself must be harmonised with Nature, *pure* morality cannot ever be actualised in itself. 'There is no *moral, perfect, actual* self-consciousness', Hegel states, only a sought-after unity of duty and reality, 'but as a *beyond* of its reality, yet a beyond that ought to be actual'.[8] In the second text, the last of Hegel's planned monographs, published in 1820 as the *Elements of the Philosophy of Right*, Hegel provides a more direct sense of 'morality' and 'ethical life'. Here, as is well known, ethics becomes bound up politically with our social and state structures, and ultimately with world history itself, so that the *Sittlichkeit* that is ethical life becomes objective rationality in the form of state laws and social institutions.[9] If the *Phenomenology* still in some sense looks back to Kant, the *Philosophy of Right* is an elaboration of the third part of Hegel's *Encyclopaedia*, first devised in 1817. More exactly it is an elaboration of the second part of that text, 'Objective Mind', before 'Absolute Mind' takes over through art, revealed religion and philosophy. *Geist* here, in fusing its subjective and objective sides, moves on its speculative journey through logic, then nature, to mind (as soul, consciousness and psychology), and then to the objective realities of law, morality and social ethics, before finally reaching its apex in the absolute spheres of art, religion and philosophy. This journey, for all its speculation, thus lends itself to a systematised Hegelianism at its most complete and most advanced stage beyond Kant.

But if we are still under the compulsion to go beyond Hegel, we cannot, as negation has now taught us, take a dialectical movement towards the ethical, however 'speculative' or 'systematic', as any more or less definitively absolute *except* by way of an absolute negation. And as we have seen, any such absolute negation would always be excessive, exceeding any 'higher' term, with its supposed stabilities of unification and universality, with its closed sense of arrival, even (or especially) if the term is 'Absolute'. Instead, the call for any ethical absolute must come from a place that understands the absolute more in terms of what William Desmond has called an 'open wholeness' – an inexhaustibility that can nevertheless find itself expressed (as in an artwork).[10] Where do we find such a place in Hegel?

Let us return to Hegel's most complete and most advanced understanding of negation as it comes to us again in the *Science of Logic*, which, chronologically, stands equidistant between the *Phenomenology* (1807) and the first edition of the *Encyclopaedia* (1817). If it is true that the *Encyclopaedia* is an elaboration and extension of the *Science of Logic*, then the art of negation should have prepared us for an inversion: any *beyond* is already within, so that any extension to the whole, as in the expanding circumference that endeavours to encompass all knowledge in the name of an encyclopaedic comprehension, is also, simultaneously, an extension of the hole at its core. That core, in the case now of an ethical extension, is the *Science of Logic*.

The *Logic*, structurally, is in reverse to the encyclopaedic *Philosophy of Mind*. Rather than moving from the subjective to the objective (from individual soul/consciousness/psychology to social law/morality/ethics), it moves, we have seen, from the objective to the subjective (from being and essence to concept/notion (*Begriff*)).[11] In the opening section, the triadic movement is from Being (*Sein*) to Determinate Being (*Dasein*) to Being-for-self (*Fürsichsein*). While in the first movement Being gains its determinateness through a dialectical process of pure being, pure nothing and pure becoming, the process begins again in a modulated form in the second: determinate being as such, *Dasein*, leads to finitude, and then on to infinitude. This then moves to Being-for-self. Negation operates fundamentally at each level of these triadic movements. But at the centre of the second process, whereby Being has become determinate, beyond its pure state, at the coordinates of the '*Da*' of the *Dasein*, and where it then moves from finitude to infinitude, we find, like a strange alien descended from another galaxy, the figure of the 'ought'. What is an ought – any ought – doing here, or doing *there* (*da*), and how can it possibly relate to the logic of Being in all its developing determinateness?

Hegel is here working out Being as determined by its finite character. To make something distinct from something else, a limit must be imposed. The pure state has no limit; but determination imposes limit, and limit marks out distinctness. But *what* in determination does the imposing? What brings limitation? We know by now the answer: negation. To render the pure state no longer, it must be negated as pure. Only then can its opposite, distinctness, arise. We have said 'imposing', but the impetus of our entire argument so far has been to reform negation under Hegel's own impetus as something 'generating'. Negation, we repeat, is never an extraneous force, but comes from within – or, as Žižek goes so far to argue, from an anterior interiority. But even in this, Being's innermost self-dividing chambers, negation yet produces something else. At this stage of the *Logic*, the negated pure state, which was without determination or distinction, brings into existence the limit of the finite state. And it is only in that state of finitude that we can say, determinatively, that some-thing exists, and exists *da* – 'there'.

But the finitude of Being, now as a singular distinction, carries with it a 'sadness'. The finite 'in its limit both *is* and *is not*'.[12] It *is*, by virtue of its new-found status as determinate; it *is not*, by virtue of the *other* that is created by the generation of the limit. By putting a boundary down, we now have two sides, and thus we bring into existence all the binary terms that we use to define these two sides: subjective/objective, inside/outside, within/beyond, self/other and so on. As the *Logic* shows, it is sometimes difficult to distinguish between the two when the limit is placed upon Being itself: 'Something has its determinate being *outside* (or, as it is also put, on the *inside*) of its limit; similarly, the other, too, because it is something, is outside it. Limit is the *middle between* the two of them in which they cease to be. They have their determinate being *beyond* each other and *beyond* their limit; the limit as the non-being of each is the other of both.'[13]

The sadness arises here in the awareness that, within all these movements operating toward and beyond determinate being, non-being comes to constitute a thing's very nature. This awareness might be seen as a permutation (or, in the system, a *premutation*) of the *Phenomenology*'s Unhappy Consciousness: 'the consciousness of self as dual natured, merely contradictory being'.[14] For what is the other of Being as determined *within itself*? It can only be non-Being. 'Finite things *are*, but their relation to themselves is that they are *negatively* self-related and in this very self-relation send themselves away beyond themselves, beyond their being.'[15] Or, more poignantly, and in a manner that Heidegger will later develop, 'the being as such of finite things is to have the germ of decease of being-within-self: the hour of their birth is the hour of their death.'[16]

Now, it is precisely out of this *coincidentia oppositorum* at the heart of Being that the ought makes its entry. The self-limitation of Being creates an internal division, by generating an externality within inwardness. But in order for these opposites to co-reside and to co-ordinate, they must continue to cancel each other out, so that the external is negated by the inwardness, as much as the inwardness is negated by the external. Both cease to be, while both cease ceasing to be, simultaneously. Such is the self-generative nature of negation: to bring something into existence, it must keep generating *itself* – a negation of negation, whose internal operation is the production of its own cogenitivity, the transversal of both subjectivity and objectivity. This transversal, as productivity, is at the same time its limit, as it cuts across both subject and object, the active power of negation and the predicative recipient of negation. This is negation's ultimate self-relation. Hegel here calls the relation to one's own limit a positive or posited negation, a relation that must therefore also, in its self, transcend the limit by negating the limit, for the limit is both the negation and the negation of the negation. This double relation, or double *self*-relation, as the *in-itself* of Being's doubled nature as being and non-being, Hegel then calls the *ought* (*Sollen*).[17]

The chief characteristic of the ought is that it is 'to be'. When something 'ought to be', its 'to be' is still only in its infinitive form. We can see this in two ways. One is negatively: it has yet to be determined finitely as an 'is', and therefore it is beyond the here and now. The compulsion to come to pass has *not yet* been made manifest as arrival. Thus in its being it has a limitation: it cannot realise itself, cannot conjugate itself. The second is positively: as an ought, something is precisely 'raised above its limitation', for the ought incurs a going beyond that something, beyond its limitation as a finite 'is'. Hegel says that these two ways are inseparable. What 'is' is not what 'ought to be'; but also, it is precisely in making what 'ought to be' what 'is' that the 'ought to be' finds its meaning, its purpose. In Hegel's terms, what determines the 'ought to be' within itself is the 'is'; and yet that 'is' is negatively related to it, since it 'is not' (it only ought to be). The very structure of 'ought' is a transcendence of itself towards an 'is', but that 'is' negates it from within. It thus both *is* and *is not*. And so it becomes a constituent feature of Being, which is caught in its own ought as the self-transcending limit of its own limitation.[18]

In all the vertiginous motions of these contradictory comings and goings, the ought is what, eventually, leads us towards the infinity ('to be') of our Being, the transcending of our limitation. But this is only possible, Hegel says, because of the negation at its heart: the breaking apart or sundering of the sameness that negation engenders, the *diremption* of identity into being and non-being, into self and other, into the 'is' and the 'ought to be', is also what turns on itself, as the negation of negation, and brings new identity between being and non-being, between self and other, between the 'is' and the 'ought to be'. But the resultant identity cannot dispense with its negative self-relation. Infinity, says Hegel, is nothing else than this generated *other* within, the perpetual going beyond the limit, the ongoing 'ought' in the 'is'. But it is also, thereby, the restored relation of the self to this other.[19] The consummation of these two opposing sides in reconciliation, in sublation of one another through the ought to the infinite, is finally what Hegel calls the Being-for-self, the *Fürsichsein* that ends the first section of the *Logic*.

Hegel does not carry forward the ought (*Sollen*) into the *Begriff* or Notion of the second half (the Subjective Logic) of the *Greater Logic*, at least not in any explicit way. It does not emerge as an *ethical* ought, much less a *Sittlichkeit*, until the more elaborated development of the *Encyclopaedia*. For what interests Hegel at this juncture is not the imperative nature of an obligatory action, required by virtue of an implicit or statutory law. The logic remains at the level of being: how does being stand before its own internal division, and how does it overcome that division to gain its freedom in infinity? The ought remains therefore a function of what Stephen Houlgate exposes as an ontological category.[20] It does not here function within a deontological category. How then are we possibly to move from the one category to the other? Is the latter category, with its questions of duty and responsible action, somehow implicit in the former category? How could we possibly move into the possibility of our being, and to beyond that possibility back to its actuality, and in doing so *obligate* ourselves toward a particular course of action?

We should know by now that Hegel, the consummate organic philosopher, does not separate the world of ontology from deontology, any more than our world of thinking can be separated from our world of action, our theory from our praxis. So too our finite existence cannot stand apart from our infinite possibility. Houlgate continues to be helpful here:

> Yet, why should finite beings, such as ourselves, seek to be anything more than beings who *should*, but never actually do, transcend their moral, epistemological, and ontological limitations? Why not just be content with remaining beings who are finite and flawed but who feel a strong obligation to better themselves? The reason, in Hegel's view, is that being confined to what we *should* be is itself a fundamental *limitation*, and by the logic of all limitation, it should itself be transcended. Finite beings not only bear within them what they should be, as well as their intrinsic limitation; they also carry the obligation and drive to become more than merely finite beings forever spanned between their limitations and their 'ought'.[21]

The Hegelian progress towards a true unity of ontology and ethics comes about through the very 'drive' suggested by Houlgate here, a drive that also becomes an obligation. But what is this drive, other than the drive we have been elucidating throughout, the drive of negation to engender within itself its own possibility? And what is this engendering, the thought that begets itself, other than the 'art' we have been formulating as a generative source of bringing into existence through the power that inheres within the act of negation? Granted, the obligation is not yet toward something specific – a specific action or practice or conformity (to a law). It is toward its own inner self, the obligation to keep obliging oneself to exceed oneself. Or it is the drive to keep driving ourselves beyond ourselves, to keep pushing our 'is' into our 'ought'. There is a fundamentally creative sense, then, to this drive. We might even say it is the drive that begets itself. And in doing so it begets its 'ought'.

But how would we *actualise* this ought, actually? Does not the ought, by definition, preclude any actualisation? This is precisely where Hegel's 'ought' ought most to inform us. For it is precisely the ought that is made actual, the ought which already has its own 'is', that negates itself as its own actuality, by which we might conceive of a way forward here. And to do this we must think of the negative impulse that allows the co-inherence of 'ought' and 'is' as itself an obligatory drive towards the possible and the creative.

But if, in a strange reversal of Emmanuel Levinas, we continue to site our 'ought' first ontologically, and not deontologically, we must, under Hegel's negative impulse, reckon with a discomfiting starting position: the site on which we ground our 'ought' is not, first and foremost, a stable site. Contrary to Levinasian thinking, in which the sameness of the self becomes the site of totality, the Hegelian site of negation at the heart of self and its being is the most unstable of all. One of the most difficult challenges in any ethics of modernity, and more acutely in any ethics of postmodernity or post-postmodernity, is how one constructs the self in relation to the ought. This is not simply the facile understanding of 'what the self ought to do'. It is more, what do we understand of the self that is supposed to carry this ought – an understanding of an ought already operating upon an ought: the ought that *ought to be*, ontologically, part of the modern self? Here the discourse of 'self-interest' has come into play, not necessarily as an ethics centred upon the self and for which all ethical decisions and actions are made in relation to promoting that self's best interest, but as a self that is coherently constituted as a sound and reliable site upon which to construct an ought. It might be suggested that virtually all modern ethics begin with this assumption of self-interest, and even Hegel himself seems to be in conformity, with his 'Being-for-self' (*Fürsichsein*), the third term in the sublation between *Sein* and *Dasein*. But as we have just seen, Hegel's self-interest is more etymologically literal in that the 'interest' of the self sits between its two opposing sides of existence, its being and non-being, its pure self and its other. This is precisely what *Fürsichsein* expresses. And in that between state, a state of transition, where we have located the ought as the coming other, there operates negation as the fundamental generative force that allows for the transition and allows the self to come into its own, as it were, to be self-interested, but through its opposite. That is to say, the self, and any self-interest, owes its existence to negativity, on what it is not. Even the later Hegel of the *Philosophy of Right* retains this fundamental premise: '"I" determines itself in so far as it is the self-reference of negativity.'[22]

Levinas' ethical project may not be so far away from this, however. As we have already hinted briefly above, Levinas in his volume entitled *Entre Nous* posits a similar sense of 'interest' for the self, as can be seen in the very title (*entre-nous* – between us) and its subtitle: *Essais sur le penser-à-l'autre* (*Thinking-of-the-Other*).[23] In the essay 'A Man-God?' of this volume, we find an understanding close to Hegel's self-reference of negativity: 'To unsay one's identity is a matter of the *I*'.[24] There is a sacrificial element of one's 'I' who stands responsibly before the other, and Levinas will use terms like hostage and victim to describe the self's position before this other, or the infinite. But in Hegel the wholly other as wholly external is already what constitutes this self, this 'I'. It is not being confronted by a proximity to otherness irreducible to our sameness. It is being confronted by that otherness as already, and necessarily, resident within us, made self-identical to us, but now not toward a totality but toward self-diremption. This is on what the early Derrida, we remember, took Levinas to task in his critique of the self/other and total/infinite binaries. The violence has already happened within. The self is already held hostage by its own negation.[25]

Those who have come after Derrida, our contemporary thinkers on Hegel in the previous chapter, from Kristeva to Malabou, have tried to show, in their way, this self-reference and self-relation of negation within. And, ethically, their implications are profound. If the ought of the self is always its internal other, then what impels the self beyond the limitation of itself is its own negation, *otherness* as such. This drive not only punctures 'self-interest' in the base sense of 'self-centredness'; it also punctures any sense of the 'self' as a stable and whole being capable of carrying out the ought in any self-containing or self-mastering or self-totalising manner. The other must necessarily define the self as part of its ethical aspect. Hegel keeps returning to the late stages of modernity – the liquid stages of modernity, as Zygmunt Bauman describes them[26] – because he became the one German Idealist to articulate fully, even 'systematically', the dirempted self that requires the other for its very being. If Levinas has gained a sense of this by working around ontology, others like Ricoeur have pursued it by addressing ontology more directly.[27] But when engendered through the negation of self, the other must go beyond even these thinkers in finding a way for self-interest to be forever blocked by its own internal operations and for any interest to be thoroughly and completely shared with the other. We repeat Kristeva's words in her *Revolution in Poetic Language*: 'negativity is the liquefying and dissolving agent that does not destroy but rather reactivates new organizations and, in that sense, affirms.'[28] If there is a totality at work in such an *inter*-ested self, it is towards a (w)hole whose oral expression betrays (like Agamben's Voice) its inner ambiguity and contradiction: the self as other, or the other as self, is built upon a hole, which its opposite must necessarily fill in order to be made whole.

The idea of giving up the self for the sake of the self is of course not new. 'He that findeth his life shall lose it: and he that loseth his life for my sake shall find it.'[29] Or more recently, Yeats' Crazy Jane says to the Bishop: 'For nothing can be sole or whole / That has not been rent.'[30] But if we were to work out an ethics from such self-negating ontology, then we would have to begin from the premise that there is no such thing as 'self-interest' in the common sense of this term. Any interest would always be beyond itself. We could not pretend that the refinement of such an ethics and its implementation on all levels of our social life would be an easy matter to realise, for self-interest is at the heart of our conception of liberal democracy. But it is apparent that it would completely alter our presuppositions about, first, the ought of the 'ought to be', second, how we would go about insisting on the ought, and third, for whose sake we would insist upon it.

The challenge here is ultimately to understand any such alteration as a creation, a gesture of an artistic impulse. For this we must impute the generative function of negation. If negation brings something new into being, even in its most interior chambers, then we cannot avoid the creative nature of the ought. The ought, we remember, breaks with the self's limitation by going beyond to an infinite. Infinity here is not the *ad infinitum* of endless succession, what Hegel famously calls 'bad infinity', or as it is sometimes rendered, 'spurious infinity',[31] but more a circular kind of infinity, which loops back upon itself to be unified again with the finite (the 'in-' prefix used in all its functions: as a privative, as directing an inward motion, as an intensive, etc.). This circle creates a space – much like the artificer's circle – out of which creation arises, or creation is made possible. It is in this sense we can understand Nancy's words that 'creation is simply Being' whose 'finitude has no measure', an infinite finitude, with 'an infinitude that consists in being its own excessive measure.'[32]

We now might argue that all of Hegel's writings, including the system that becomes known as Hegelianism, are imbued with this kind of excessive poetical character in which negation is continually re-creating, or re-inscribing, the site from which any supposed third term arises, or any dialectical movement is maintained. A fourth, a fifth term is always waiting to emerge. The narrativisation of this infinity is of course most poetical in *Phenomenology of Spirit*, but it appears throughout the Hegelian corpus, even as as far as the *Philosophy of Right*: 'infinity as self-referring negativity, *this ultimate source of all activity, life, and consciousness.*'[33] If the ought itself is a creative source, then we cannot help but rethink ethics from an understanding of art, an art of negation, which not only disinters Nietzsche, but more, places the artist back into the responsibility of an ethical domain, no longer as the singular Nietzschean *Übermensch*, but as the social creature she always was, now bound to dirempting herself by means, we might say, of the demands – the ought – of her art. Here the beginning works of Kristeva, Nancy, Agamben and Malabou initiate a theorising of the generative *poesis* of Hegelian negation. As we saw in Agamben's *The Man Without Content*, the artist becomes, negatively, the one 'who has no other identity than a perpetual emerging out of the nothingness of expression and no other ground than this incomprehensible station on this side of himself'.[34] But in Nancy, art gains, positively, a new sense, or sensibility, of 'sense' – a unity or identity of sensation (the finite stuff of art) with meaning and understanding (infinity or ideality that art strives for beyond its materiality) – 'sense' in all its senses. To *make sense*, then, would be the supreme task of the artist. And this making, as a *speculative* making, would have to make a space for nothing, in order to allow the other to come forth from within (within the self, as the self). Art would thus have to make sense of the ought and in multiple senses: it would have to actualise the ought in the material world of our senses; it would have to give meaning to the ought; it would thus have to *make the ought* as the hermeneutically interpreted other; and the ought would have to make art as the other that compels creation.

In combining the dirempted self with this creative ought, we might say that both self and other would be displaced, would be re-created and *sublated*, into a 'we': neither subject nor object (as in Kristeva's abjection), but a plurality, even within the singular. This of course would be to follow Nancy's path of 'being singular plural': 'Being cannot *be* anything but being-with-one-another, circulating in the *with* and as the *with* of this singularly plural coexistence'.[35] For us, it is circulating in the *of* of cogenitivity. But it also suggests, along with Agamben, that any art is also displaced from the singularity of the artist and must move outwards towards the social in a more inclusive hermeneutics of neither subject nor object. Again Nancy's words: 'The "end of art" is always the beginning of its plurality.'[36]

In following these pluralities we would also displace the concept and the highly developed discourse of human rights, upon which modernity, and all its present politics, are so fundamentally grounded. In this sense, Hegel's own concept of 'right', as articulated in his last full work, is exceeded and gone beyond. For it is no longer that personal right and its accompanying subjective freedom are consummated in the institutions of the State – an idea that has worried many prior to the twentieth century, and repulsed most during and after the twentieth century. It is that 'right' succumbs to the 'ought' – not as the imperative norms of morality or the sovereign politics of *Sittlichkeit*, but as personal freedoms limited by the other towards a new possibility, freedoms, that is, made possible by the necessity of a certain poetry of negation: or, and this amounts to the same thing, as the affirmation of an ideal beyond the self-interested claim of 'rights', and towards a re-creation of the self as 'we'. This is not the ideal 'we' of idealism, the absolutised *Geist*, the spirit of totalising philosophy. It is the actualised 'we' recreated, following Malabou, as a future reality already taking place here and now.

If, then, we are to trace out the contours of an ethics of negation that becomes a poetics of negation, we must retain Malabou's sense of the future of Hegel, which can be correlated to Ricoeur's definition of ethics: an optative mode of living well in distinction to an imperative mode of obligation.[37] Hegel's ought, as reconceived, is ultimately to be seen in the optative mood, where desire and future hope go beyond the obligatory laws and customs as encrusted in the present systems of our beliefs and institutions. But then, as Nancy has pointed out, Hegel's system is always driven by the optative mood: 'Self-consciousness is essentially desire, because it is consciousness *of self* as and out of its consciousness *of the other*.'[38] But this drive is also, in the most Hegelian fashion, towards the indicative, towards the here and now. And it is the optative as indicative, the 'may it be' as both 'ought' and 'is', Hegel as himself beyond himself, that we now need to think on and act on most acutely, and most creatively.

CONCLUSION
Art-Religion-Philosophy Re-formed

In his dialectical lyric *Fear and Trembling*, at the beginning of Problema I, Søren Kierkegaard, under the name of Johannes *de silentio*, writes of faith:

> For faith is just this paradox, that the single individual is higher than the universal, though in such a way, be it noted, that the movement is repeated, that is, that, having been the universal, the single individual now sets himself apart as the particular above the universal. If that is not faith, then Abraham is done for and faith has never existed in the world, just because it has always existed. For if the ethical life is the highest and nothing incommensurable is left over in man, except the sense of what is evil, i.e. the single individual who is to be expressed in the universal, then one needs no other categories than those of the Greek philosophers, or whatever can be logically deduced from them. This is something Hegel, who has after all made some study of the Greeks, ought not to have kept quiet about.[1]

In Kierkegaard's inverted Hegelianism – an existential dialectic, whereby what comes to self-expression is not world *Geist*, but the particularised self in its 'shudder of thought' – the universal gives way to the passion of the individual, expressed here in faith, the 'highest' of passions. Abraham is a particular man of faith, not a universal representative or typology of religious obedience. He is called a 'knight of faith', but this knight does not and cannot stand for Everyone in their struggle for religious or spiritual authentication. The paradox, however, is that he must go through the universal, as ethical, in order to reach that faith.

He must understand the ethical as an obligation for everyone, including himself, but he must surpass it, by way of the absurd – for there is no reason that can guide him – and decide for the wholly unreasonable, even the wholly unethical. If he stays in the ethical, as it is universally and teleologically mediated, then he is, as Johannes *de silentio* says, 'done for'. He could only be designated a murderer, for his actions would contravene the universal requirements for family obligation and social harmony. But in choosing to surpass the ethical in faith, Abraham finds his truest justification. There can be no reasonable explanation for his decision. If one is looking for an archetypal reason, for an *archetype* as such, one needs to go no further than the Greek philosophers, whose categories provide the kind of *arche* upon which any type of sound ethical prescription might be constructed. But Greek metaphysics does not hold in Abraham's case. And Hegel, says Johannes, should have known this. Or at least, he should not have kept silent about it. Thus the paradox of a man *de silentio* who voices the true nature of the case.

But Hegel, contra Johannes, contra Kierkegaard, does not remain silent. Or, he speaks his silence forcefully in the name of negation. For negation is just this paradox, that the single individual, as self, cannot be higher than anything else, any more than the universal can be higher than anything else, but both must give in to the passion that is negation, because both have *already given in* to the passion that is negation, a negation that works first upon itself, in order to allow both the self and the universal to emerge into being. This negation, as silence, is neither commensurable nor incommensurable with what it originates. It exceeds the question of commensurability, because in its negativity there is nothing to measure, and nothing to measure up to. The Greeks, in their Parmenidean aura, shunned this irrationality. But Hegel, the supposed arch-metaphysician, did not. And this is why, just as Kierkegaard returned to Abraham, we return, now again, to Hegel.

Without this negation, without this *faith in* negation, Hegel is, we might say, done for. But Hegel, and his Hegelianisms, have long been done for, in their various ways. As an idealist, as a universalist, as the apex of Enlightenment reason, Hegel and the Hegelian position, however they are constructed, have always gone to ruin. Even the inverted Hegelianisms, dialectical materialism, say, have not been able to survive. So what keeps Hegel returning, and returning anew? What keeps saving Hegel in the withdrawal of his (self's) destruction, or in the destruction of his destruction, or in the recantation of that destruction? The paradox that is negation.

Following Johannes *de silentio*, we have tried to bring the silent and hidden nature of this negation out into the open. Johannes says later in Problema III that concealment is absolutely necessary to gain and retain the faith that

Abraham achieved. 'Unless there is a concealment which has its basis in the single individual's being higher than the universal, then Abraham's conduct cannot be defended, since he disregarded the intermediate ethical considerations.'[2] But in wilful contradiction, Johannes discloses the concealment, lays it bare for our consideration. To re-voice Derrida: 'Such a silence takes over his whole discourse.'[3] And once again, Kierkegaard implicates Hegel, but again for being on the wrong side of the tracks: 'Hegelian philosophy assumes there is no justified concealment, no justified incommensurability.'[4]

But Hegel, contra Johannes, contra Kierkegaard, does not assume this. Negation does not only assume concealment – it enacts it. And in that very enactment, it comes into its own as manifest, as unconcealed. This is the great trick of negation. When negation turns on itself in self-negation, it is most manifestly itself. When Hegel is done for, he makes himself incommensurable with himself, but in that very incommensurability he is most himself, most measured against himself. Or to be more Hegelian: he is most immediate to himself when he mediates himself through negation, which quickly makes immediate the mediacy.

Kierkegaard, through Johannes, gives Hegel at least this much: Hegel saw the first immediacy as aesthetic. Here, he says, 'the Hegelian philosophy may well be right.'[5] Though we can say Kierkegaard himself is right only with much qualification,[6] nevertheless, if we take the movement towards the highest passion as ending in faith, then the aesthetic, as the starting point of immediacy – that is, when the powers of reason have not yet infiltrated experience – must for Kierkegaard be left behind. This much he agrees with Hegel, even if what aesthetics is left behind *for* is not religion and then philosophy, but ethics and then religion. In the highest passion of faith, Abraham is not done for: rather, aesthetics is done for, and ethics is done for, on the strength of the Absurd. In the highest reaches of Absolute Knowledge and Spirit, we know from Hegel that art is done for: art, far removed 'from being the highest form of spirit, acquires its real ratification only in philosophy'.[7] But in the 'highest' passion of negation, which does in the distinction between 'high' and 'low', Hegel is not done for: rather, Hegel is able to return anew, in the cogenitivity that is the 'return of negation' and at the same time is the 'negation of Hegel'. And neither is art done for, for in the cogenitive dynamic of the 'art of negation' we also find the 'origination of art', and we do so *in all* areas, in all so-called phases and development, of existence.

Let us then re-emphasise what we have been saying throughout. If, in the Hegelian system, and according to the readings suggested by the *Aesthetics*' now infamous Introduction, art is done for, because art must be surpassed, first by religion, and second by philosophy, then in the *negation of Hegel*, whereby negation becomes an *art*, so too is religion done for, and philosophy done for. If negation is truly an originary power that, in breaking asunder, allows the new to emerge, then neither art nor religion nor philosophy can properly stand in their received senses and conventional formations, or even in their derivatives (aesthetics, ethics, etc.). All are pulverised by negation. And therefore all are *art* in its most originary sense.

Here we can invert Kierkegaard. If, under the aesthetic or the ethical as the highest, faith has never existed just because it has always existed – a general or universal faith is for Kierkegaard the highest of contradictions – then we can say that art can no longer exist *just because it has always existed*. And the contradiction here, in all its extremity, is *precisely the point*. The art of negation becomes the negation of art precisely because art now infiltrates all manner of our being, as negation. This cannot be to say that all life becomes *aesthetic*. The reflective nature of aesthetics is still relegated to specific moments and specific gestures, as an always and only mediated relation to beauty and creative works. But the art of negation is the immediate force of the most interior act of origination, whose own mediation is always being rendered immediate by its self-sundering operation, even if that operation itself requires mediation. The contradictions at play here are what allow art its potency and its ubiquity. If art, in any conventional sense, is done for, it is because it is reconceptualised beyond the *fait accompli* upon which our aesthetic gaze is encouraged to rest. It is reconceptualised beyond the concept: it is the Concept, as *Begriff*, rent within, negated by its own devices, to return to its immediate origination. Thought that begets itself. If art is done for, in the passive state of ruination, it is ruined for its very opposite, an art that *does for*, the active and even ethical movement of a creative urgency. It is in this sense that Hamacher can claim: 'in order to be art, art cannot simply be itself; it must also be the art of the dissolution of art.'[8] Art must dissolve its own completion, its own self-imposed limits, and re-do itself out of its own abolishment and desert sands. Only then can it become, beyond Concept, 'a release of matter without contour and of thinking without schema, as a dispatch in which *with* art something other *than* art, something other *as* art, is promised and exposed.'[9] And this something other *as art* becomes, among other things, religion and philosophy.

We are tempted here to write the *Phenomenology*'s art-religion as art-philosophy, and then, by extension, as art-religion-philosophy. But the concatenation of these three terms risks the impression of a merged stage of positive and privileged domains. Art does not now, under negation's impetus, simply merge with religion and philosophy in an equilateral manner. All three undergo a mutual dissolution, out of which emerges a wholly new configuration for each. The art of negation privileges the first term only insofar as it is not the first term in its accomplished, 'fine' or aesthetic sense. It is rather the term as the undercurrent of negativity that is always in force by allowing the possibility of any configuration to arise into view. Philosophy-religion-art. Religion-philosophy-art. Art-philosophy-religion-ethics. And so on.

*

Anton Chekhov's short story 'The Black Monk', published in 1894, opens with a young academic named Kovrin, 'master of arts', in a state of mental and physical breakdown. Having been successful as a scholar – mainly 'in philosophy' – he is advised on medical grounds to retreat to the country for recuperation, where he ends up at the estate of his former guardian and tutor, a renowned horticulturist. There, in the frequent company of the horticulturist's daughter, Kovrin begins to rejuvenate, even though his excessive habits – restlessness by day, sleeplessness by night and a proclivity for much wine – do not abate. One evening, upon listening to the young daughter sing of a young maiden with a febrile imagination, who hears sacred sounds too lofty for mortal comprehension, Kovrin recounts to the daughter a legend he had once heard of a monk who, dressed in black, wandered the deserts of Syria or Arabia, and then began to appear, mirage-like, in disparate places around the globe, 'never able to find the right conditions which would allow him to evaporate'.[10] The legend told of the monk's return after a thousand years, when he would start appearing again to people, a period that had just now elapsed. And sure enough, in subsequent days, Kovrin meets this monk in a series of visionary encounters.

We have here a typical modern characterisation not of the scholar or philosopher, but of the artist: prone to excess, prone to nervous collapse, prone to visionary flights of fancy. And when a certain egoism overtakes the young man in the wake of these encounters, leading to his eventual demise, that characterisation appears complete: the self-destructive 'genius' who can only maintain his genius in the intensities of a hallucinatory madness that all too quickly consumes his creative energy and his life force.

But who is this black monk populating his visions? He is a desert figure, whose black habit marks him out against the blanched backdrop of his surroundings. This blackness might suggest an infernal figure, of the Mephisphelean kind, but his own message would appear to contradict any kind of symbolism along these lines, any equation with the negative, even if merely a 'bad negative'. For the black monk brings, like many a desert figure, a seemingly divine message. '"You are one of a few people who can genuinely be called one of God's chosen people",' he tells Kovrin. '"You serve eternal truth. Your thoughts and intentions, your astounding scholarship and your whole life have a heavenly, celestial bearing, since they are dedicated to all that is rational and beautiful, that is, everything which is eternal".' And he goes further:

> Mankind would be nothing without those of you who serve the highest ideals, living consciously and freely; if mankind followed its natural course it would have to wait a long time for the culmination of its existence on earth. You are bringing mankind into the kingdom of eternal truth several thousand years ahead of its time, and that is your greatest contribution. You are the incarnation of God's blessing, which is immanent in humanity.[11]

We could read this in the great modern tradition of art's aggrandised role: it is the visionary as artist who has become the new priestly caste. We could even read this by means of a Hegelian modulation: it is the artist (with an eye on philosophy) who makes immanent the God of heaven *in humanity*. And the black monk is this vision made manifest through the power of imagination. The monk would then represent the form and content of art, religion and/or philosophy: what is envisioned is precisely the power to envision.

But here too we need to be wary of an over-hasty interpretation. The monk, for all his heavenly tidings, does not secure the blessed future. For it is precisely in believing this message – that he, as visionary, is one of God's chosen – that the artist/philosopher begins his demise. Kovrin may have had a productive resurgence upon the first encounters with his vision, but, after marrying the horticulturist's daughter, and accepting a professorial chair within the academy, his creative and scholarly powers soon wane, his daily existence becomes increasingly miserable, his marriage eventually disintegrates, and he scorns the very guardian who first provided him care and retreat. He becomes the modern 'tragic' figure, except there is no real tragedy in his situation or his actions.

Kierkegaard would have had troubles with this story. In Kovrin he would have seen only a failed aesthetic, who disclosed his vision when he should have concealed it (in revealing his encounters with the black monk to his wife he only confirms his madness to her), and who concealed his vision when he should have disclosed it (he never reveals why he spurns the horticulturist and his daughter and why he leaves them to their own demise). He is certainly no tragic hero, since the ethical is absent from the situation (the monk may offer an exalted purpose for Kovrin's existence, but it cannot be ethically fulfilled, since it requires a *departure* from reality and rationality, the madness of the visionary). Nor is there any faith on the strength of the Absurd, for the monk is only the *artist's* absurdity, which holds out the promise of an eternal that can never be found within. What Kovrin would have faith *in* is his own visionary power, and for Kierkegaard, as much as for Hegel, such is a passion deeply misplaced. The story bears this out: Kovrin's genius implodes in upon itself, taking others with it. The aesthetic, as it pulls in the philosophical, here shows its impotence.

In all the ambiguities of this story, which do not allow even an anti-heroic reading, what then do we make of the black monk? Is he angelic or demonic? Is he saviour or destroyer? Let us look again to his initial conversation with Kovrin. After the above quotation, in which is announced a spiritual calling, Kovrin asks, 'But what is the goal of eternal life?' And the answer he receives is this: 'The experience of delight, like in any other life. True experience of delight comes from consciousness, and eternal life presents innumerable, endless sources for consciousness; that is what is meant by "In my Father's house there are many dwelling places".'[12] In this response, which rewrites the scriptural promise in wholly modern terms (desire, consciousness), we might find the nucleus of the story's enigmatic nature, its radical indeterminateness that does not allow us to place either Kovrin or his vision in any definable category. (Who here is the protagonist, who the antagonist?) For if, in the achievement of the highest passion or calling, one is not consummated in the singularity of his or her consciousness – and let us remain here with this translation, 'consciousness'[13] – but instead is immersed in, only to be diffused across, a multiplicity, not of consciousness itself, but of *sources* of consciousness, we gain a sense of why Kovrin cannot fulfil this calling, cannot sustain his life. The aesthetic, the philosopher, who believes in the ought of his self-sustaining knowledge as a single consciousness singled out for necessary greatness, will be irrupted from within. Or the one who sees himself divinely elected to advance humanity ends up destroying that humanity, himself included. Even the individual reformer, acknowledges Kierkegaard, 'must see to it that he himself falls'.[14] The black monk, then, is not the interior vision of the artist *par excellence*, nor the calling to a higher idealised purpose, either religious or philosophical, but the force of a negativity that opens up the space of origination, of manifold sources for the consciousness, through the utter incommensurability of its invocation. What consigns Kovrin to the mediocrity that ushers in his demise is not the abandonment of the monk's calling (through marriage, through business concerns, through daily practicalities), but his inability to grasp what that calling truly entailed. It is the abandonment of his self that is most needed, and that only truly comes with his dying breath, which forms on his lips a 'beatific smile'.

Chekhov gives us a tale here of art disrupting itself as art. The work is, assuredly, a finely crafted narrative, and an aesthetic wonder on many levels, as are so many of Chekhov's short stories. But the ambiguous nature of its central vision – the black monk whose call to genius in eternal life is also a call beyond any individual genius, a call to the innumerable sources of negation's power – takes art to a different space. This space is the art of negation, an internal diremption that, in the end, leads to an excess of consciousness beyond any one artist, any one philosopher, any one genius. It is the space that, even in their modernist or postmodernist singularity, so many artists of the next century will actively pursue – alongside, as we have seen, several philosophers. It is the space that also, even here in this story, engulfs religion.

*

If the blackest monk in our contemporary world is Thomas Altizer, it is because, as a self-professed theologian, who has wandered long in the deserts of death of God theology, Altizer has most consistently seen the necessary centrality of Hegel's negation, and understood this negation as an art. In all his understanding of nothing as Nothing, and abyss as Abyss, Altizer not only places Hegel in the centre, at their origin, and he not only places Kierkegaard alongside this Hegel, in an inversion of passion as subjective *Angst*, but he also places there origin itself, as both origination and as transfiguration. Arrayed with the hooded habit of Chekhov's monk, Altizer announces in *Godhead and the Nothing*: 'Visions of transfiguration have never been more comprehensive than they are in full modernity, and if these are inseparable from visions of an absolute abyss, that is the very abyss offering the possibility of an absolute transfiguration, even if it can only be realized in the deepest depths of darkness and abyss.'[15] Just as Chekhov helps to bring in a new modernism, so too this transfiguration is always toward a new creation, to the point where creation and transfiguration become inseparable, and indeed identical. Hence Altizer's ceaseless fixation upon both genesis and transfiguration. In the bi-directionality (going back to origin, going forward to change) and in the paradox (the endless movement of fixation), a motion is initiated that is always the circular cogenitive movement of a 'genesis of nothing' and a 'transfiguration of nothingness', for it is only in the genesis and transfiguration *of nothingness* that cogenitivity and paradox are possible. At such a point, an absolute Yes is inseparable from an absolute No, the realisation of which, Altizer tells us, is 'not fully imaginatively born until *Paradise Lost*, and not fully born philosophically until the *Phenomenology of Spirit*'.[16] One can find this genesis and transfiguration repeated and reiterated – *reborn* and *transfigured* – throughout all of Altizer's writing, a *transgenesis* that occurs so often in the name of Hegel, and always as Hegel's negation.

More radically, as the title *Godhead and the Nothing* indicates, this movement occurs *within* the Godhead itself. As we saw in Derrida's *Glas* and later in Žižek, the death of God is not merely the death, through the crucifixion, of the human Christ dialectically subsumed back into the Godhead, maintaining His redemptive power. The movement of nothing, or the act of negation, takes place within the already divided Godhead as triune: Father, Son and Spirit. This is the real crux of the death of God theology. In Altizer's most recent work, *The Apocalyptic Trinity*, this crux is at its most assertive. The Trinity manifests itself, not only *in* an imaginative or philosophical drama such as that envisioned by a Milton or a Hegel, but more importantly *as* an internal drama unto itself, a drama of self-negation. Each figure within the triune structure, whether as person or mode, enacts a self-negation, but a self-negation that is the other within (the Father self-emptied as Son, the Son self-emptied as Spirit, etc.). In Hegel, this enactment becomes thinking itself, so that, as we have tried to show repeatedly, following the *Phenomenology*'s own lead, thinking itself is an internal drama, one that is built upon a triadic movement that becomes, for Altizer, the very movement of the Trinity. Thus, 'we can understand the Trinity itself as being profoundly dialectical, so that it is quite naturally a center of Hegel's thinking and perhaps a center of all Western Christian dialectical thinking.'[17] But this is not merely to suggest that Hegel was deeply influenced by Christian theology; what Altizer is really saying is that Christian theology is now and by necessity deeply influenced by Hegel, 'the deepest trinitarian thinker in the modern if not in any world, a thinker who created a trinary logic, and a logic revolving about the absolute self-emptying or self-negation of that Godhead which Hegel knows as Absolute Idea or Absolute Spirit'.[18] If the Trinity was in-itself in orthodox Christianity, it was for-itself in Hegel, and becomes in-and-for-itself in the death of God thinking of late modernity.

But if the Trinity is an internal drama, a drama that plays itself out through the art of negation, that drama is thereby a tragedy. Tragedy is crucially religious, Altizer reminds us, where 'the origins of tragedy lie in a primordial ritual of atonement or sacrifice, one whose sacrificial victim is deity itself', so that our 'purest tragedies are enactments of sacrificial deity'.[19] The tragedy of the Trinity is precisely its internal enactment of its self-sacrifice, its trinary kenosis through self-negation, which produces a comprehensive *agon* within. If the theological term *perichoresis* (interpenetration) has any currency for Altizer, it is as a perichoresis of passion within the Godhead. The dramatic movement towards death that we associate with Calvary is the dramatic movement happening throughout the whole of the Trinity, leading to the death and transfiguration of *all* within the triune God.

But this transfiguration radically transfigures tragedy itself. Tragedy, in this post-Hegelian world, cannot remain the purely tragic, but always succumbs to its opposite. 'Now it is not insignificant that of all our major cultural creations it is tragedy that we least understand, or perhaps tragedy and comedy together, and just as comedy is not alien to tragedy, our greatest late modern comedies are tragic comedies, as above all realized by Chekhov and Beckett. Thereby tragedy itself is transfigured.'[20] Is this why we have such difficulties interpreting 'The Black Monk', because it is neither tragic nor comic, but both tragic and comic at once, the destructive death that gives birth to the beatific smile, the *agon* and *Angst* overcome by the self-sundering that bore it?

The tragic Trinity, which nevertheless gives birth to a new creation and a perpetual transfiguration, even if only in the excess of itself, brings together here art, religion and philosophy in a manner we have not yet seen: a *theological* reading that requires, ineluctably, the imagination of art as much as it does the conceptualisation of philosophy to drive it forward. To read Altizer is to engage with all three simultaneously in their own unique perichoresis or circumincession. But his work also suggests to us that this privileged Hegelian trio – art, religion and philosophy – likewise has its own internal drama, its own Trinitarian movement, and if one is done for, *they all are done for*. If the self-negation is a kenotic drama between each mode of being, of acting, of producing, of conceptualising, then each individually is self-emptied, each is negated, each is a thing of the past. And yet, each must be reborn, and be reborn from within – and not apart from – their internal dependence, or what we might now call a *trigenitivity*. The art of religion of philosophy. A tri-fold prepositional relation that becomes propositional across its three parts. 'Tri-fold' *in the very least*, we should add – for we cannot yet predict what new births might await us beyond any trinary scheme, whether Christian, Hegelian, Kierkegaardian, Marxist and so on. If Altizer is right that something original (art, Christianity, Hegel…) is only renewed in truly radical movements,[21] a trinary scheme awaits to be revolutionised by the art of negation. Which is to say, in Hegel's case, by his own *Aufhebung* through which we return to him anew.

*

> At the same time, along with the free, self-conscious creature, a whole *world* comes to the fore – out of nothing – the sole true and conceivable *creation out of nothing*.
> …
> The philosopher must possess as much aesthetic force as the poet… The philosophy of spirit is an aesthetic philosophy…
> …
> Poesy will thereby attain a higher dignity; in the end she will again become what she was in the beginning – *the instructress of {history} humanity;* for there will no longer be any philosophy,

any history; the poetic art alone will survive all the other sciences and arts.

At the same time we so often hear that the great mass of human beings must have a *sensuous religion*. Not only the masses but also the philosopher needs it. Monotheism of reason and of the heart, polytheism of the imagination and art, that is what we need!

...

... mythology must become philosophical, and the people rational, while philosophy must become mythological, in order to make the philosophers sensuous. Then eternal unity will prevail among us... A higher spirit, sent from heaven, will have to found this new religion among us; it will be the very last and the grandest of the works of humanity.[22]

Who wrote these words? By all accounts, Hegel, and a young Hegel, in 1797. That is to say, it is universally agreed that Hegel was the one who picked up the pen and committed to paper these words, as fragments of a fragment, now known as *Das älteste Systemprogramm des deutschen Idealismus* (*The Oldest Programme toward a System in/of German Idealism*). But are they Hegel's *own* words, that is, are they a true representation of the young Hegel's mind at that time of his life when he had just left his job as a house-tutor in Berne and moved to Frankfurt, still years before his career was properly launched (i.e. during the Jena period, 1801–1807)? What is the relationship between Hegel's written words of this famous fragment and Hegel's own interior *Geist* of this period? This question has been the fodder of much academic debate concerning the authorship of the *Systemprogramm*. Most (not all) agree that the Programme was a collaborative effort conceived and worked out between the three theological roommates at the Protestant seminary at Tübingen's university: Hölderlin, Hegel and the young upstart Schelling. But there is much division about who actually articulated the words of the fragment after the three had left seminary and gone to separate posts. And much argument has been made for each of the three, even if the general attribution of authorship, not just the penmanship, might still default to Hegel.[23] Here then we have another crucial trinity, a trinity of German Idealism, which manifests itself only in a fragment (in fact, a torn piece of foolscap) and with the same ambiguity of internal relation that Altizer has pointed us to in the doctrine of the Trinity.

David Farrell Krell opens his expansive argument, *The Tragic Absolute: German Idealism and the Languishing of God* (2005), with an exposition and translation of the *Systemprogramm* fragment, followed by a commentary. The philological debates, he shows us, make it fundamentally *unclear* about who the true author may have been, even if compelling cases might be put forward for each in respect of certain passages of the fragment. What emerges most clearly is that, in the generation of the Programme, collaboration stands above influence. It seems three minds bonded together in an attempt to remake the fundamental relationship between art, religion and philosophy, and thus between humanity and divinity.

Krell, whose ultimate concern is with the question of tragedy, in a monumental text 'at the crossroads of metaphysics and aesthetics, ontology and literary criticism',[24] wonders about our obsessions with trying to pin the authorship down. 'Why', he asks, 'the *competition* among the disciples of Schelling, Hölderlin, and Hegel, when these three allowed their ambitions to be yoked to a common task, a shared historical-epochal vision and mission?'[25] Why this need, this necessity for singular origin, as stemming from a singular consciousness? Why can we not allow mutual creation in this case? What dictates against a perichoresis of young reactive seminarian thought? For Krell, the *Systemprogramm* is a fully trinitarian expression in the sense that Altizer relays it.

As Krell goes on to develop a notion of the tragic at work in German Idealism and its 'languishing' God, we might see the importance of his beginning with the *Systemprogramm* in this way: the absolute, even in the sense of an absolute singular consciousness, opens itself up to its own condemnation, its own inevitable fall, precisely by being open to a multiplicity of consciousnesses. (Agamben: 'Philosophy, in its search for another voice and another death, is presented, precisely, as both a return to and a surpassing of tragic knowledge.'[26]) If German Idealism is both a system and a programme, combined in the most Germanic of fashions, *Systemprogramm*, it had in its very core the dissolution or dissipation of its very unity and coherence. But that split core is precisely its greatest strength. Multiple cognitions, multiple consciousnesses. The artist, the theologian, the philosopher all joined in a ring, which forms not only a dance of death, the tragi-comic response, but the artificer's circle, in which new creations arise.

If we return to Hegel anew, then following Krell we return to the origin when Hegel was not Hegel, or not *only* Hegel. *Creatio ex nihilo*, aesthetic philosophy, sensuous religion, unity of *mythos* and *logos* – these may all be the hallmark of a young Hegel's mind, but only as mediated through others, and mediated *as* others. That disconnect between the written script, the material expression of handwriting and the possession of the concept, the production of one's own mind, is here, at the origin, made ambiguous in the same way that Derrida tears through the tympanic membrane separating the exterior sound from the interior thought at the threshold of the inner ear. Where does the outwardly expressed Hegel stop and the inwardly impressed Hegel begin? The question, beyond the

System that became Hegel's, a beyond that is now a before, is answered across multiple consciousnesses. Hegel stops and starts with others. He has his origins in that which is not himself. At the beginning he enacts a trinitarian movement, by which, in the end, even he will be undone, done for. For he must break with, negate, his fellow travellers to become the Hegel we know, beyond the fragmentation of a system towards the apotheosis of a System. But that negation requires his own as self-negation. To break from the early *Systemprogramm* is to move from *creatio ex nihilo*, where his collaborators are negated, to *creatio qua nihilo*, where all three succumb to their own negation, *qua* negation. This ultimately is the tragic absolute Krell wishes to unveil.

But this negation opens up a new Hegel, just as it opens up the System or the Programme. It does not just retain the original three; it exposes them all to the centrifugal force of the art of religion of philosophy that explodes beyond three. Under such a movement, Krell is fully justified to claim, '*per impossible*', that the real author is yet a fourth figure – Nietzsche.[27] And his anachronistic authorship is precisely possible in the same way that at the origin, before Hegel overturned the philosophical world with the publication of *The Phenomenology of Spirit* in 1807, Hegel was not Hegel, not fully present, and yet, in the paradox of self-negation, was more present than even he could have possibly imagined.

*

Why, given the controversy surrounding the *Systemprogramm* authorship, did not Hegel ever say anything about the writing himself? Why did he, like Kierkegaard's Abraham, remain silent on the matter? If we return to the young Hegel's poem, sent to his friend Hölderlin only a year before in 1796, the 'Eleusis' Agamben drew our attention to earlier, we might receive our whisper. If there Hegel is, as Agamben claims, the guardian of the Eleusian mystery, it is because he understands that one guards the ineffable through the manifested acts and rituals dedicated to it. As Kierkegaard later saw, and Derrida after him, if one wants to preserve and honour a necessary silence, one must develop a highly articulated language for the purpose, a language that is as much enacted as spoken, and maybe more so.

> And so you did not live on their lips.
> Their life honoured you. And you live still in their acts.[28]

Hegel is *de silentio* about the collaborative authorship because perhaps, secretly, he knows the madness that can derive, as in 'The Black Monk', from the eternal delight of multiple consciousnesses. Nietzsche knew this all too well, at precisely the time Chekhov's story was penned. And so, eventually, did Hegel's friend Hölderlin. Silence kept Hegel from such madness, as it did Abraham. But, in the name of a negation, it did not the System. In articulating an art of negation, the System collapses in on itself. It is done for. And this, in an inversion of the inversion of Kierkegaard, is its greatest strength.

At the end of *Fear and Trembling*, Kierkegaard recounts that famous Heraclitean gnome: one cannot step into the same river twice. But a disciple was said to have added, in the name of improvement, that one cannot do it even once! But the improvement, Kierkegaard tells us, led to its opposite, an Eleatic doctrine denying movement, even though the disciple wanted to go further, not back to what Heraclitus had abandoned.[29] In the inversion of this parable we find the present engagement with our histories of Hegelianisms: to go further, as disciples, in the System so comprehensively set up by the mature Hegel, we go back to what that System abandoned, because it had to: an art of negation. But if the abandonment is – already, necessarily, inevitably – part of the System, if the art of negation drives all its movement, and has done so all along, we will have gone further by going back. We will have returned anew.

'One must go further, one must go further.' This must also be our call to the future, retrospectively.

Notes

INTRODUCTION: RETURNING ANEW

1. Frederick C. Beiser, 'Introduction: The Puzzling Hegel Renaissance', in *The Cambridge Companion to Hegel and Nineteenth Century Philosophy*, ed. Frederick C. Beiser (Cambridge: Cambridge University Press, 2008), pp. 1–14.
2. We recall Marx's famous gloss on Hegel at the opening of *The 18th Brumaire of Louis Bonaparte*: 'Hegel remarks somewhere that all facts and personages of great importance in world history occur, as it were, twice. He forgot to add: the first time as tragedy, the second as farce' (New York: International Publishers, 1963, p. 15). That is, repetition is never the same. And though revolutions 'conjure up the spirits of the past to their service' (ibid.), such awakening serves 'the purpose of glorifying the new struggles, not parodying the old... of finding once more the spirit of revolution, not of making its ghost walk about again' – p. 17. Hence: 'The social revolution of the nineteenth century cannot draw its poetry from the past, but only from the future. It cannot begin with itself before it has stripped off all superstition in regard to the past. Earlier revolutions required recollections of the past world history in order to drug themselves concerning their own content. In order to arrive at its own content, the revolution of the nineteenth century must let the dead bury their dead' – p. 18.
3. What is being *conserved* in this political agenda is only the belief that liberal democracy is inexorably progressive, that is, politically progressivist and philosophically teleological, and not anything substantive that society could or should return back to (the 'neo-' pointing to the renewing of the future more than to any new way of envisioning the past).
4. We should recall that Ricoeur understood the hermeneutical relation of explanation and understanding, of sense (*Sinn*) and reference (*Bedeutung*), of participation and distanciation, of textual and reader identity, very much in terms of a dialectical process. One 'appropriates' the text, then, only when a dialectical fusion and sublation (Gadamer's fusion of horizons) occurs between the two sides. See for example Paul Ricoeur, *Hermeneutics and the Human Sciences*, ed. and trans. John B. Thompson (Cambridge: Cambridge University Press, 1981): 'The dialectic of distanciation and appropriation is the final figure which the dialectic of explanation and understanding must assume' – p. 183. Again, in terms of Hegel, Hegel is needed ('with') to read Hegel ('within').
5. The question of re-raising the spirit of Hegel, the ghost of Hegel, is the same as that in Hamlet, who questions the raising not only of his dead father and 'poor Yorick', but of the very nothing of his own being. As Andrew Cutrofello has colourfully shown us, Hamlet's nihilism haunts Hegel as much as Hegel's negation haunts Hamlet. In Hegel's own reference to Hamlet in the notorious phrenology section of the *Phenomenology of*

Spirit (spirit's relation to the bone), Cutrofello writes: 'The question of what it means to sublate the dead troubles spirit for the rest of Hegel's *Phenomenology*' – 'Hamlet's Nihilism', in *The Movement of Nothingness: Trust in the Emptiness of Time*, eds Daniel Price and Ryan Johnson (Aurora: The Davies Group Publishers, 2013), p. 126. See esp. pp. 123–126.

6   Fukuyama's adaptation of Kojève's reading of Hegel in *The End of History and the Last Man* (London: Penguin, 1992) became a mandarin text for neoconservative politics in America, despite Fukuyama's subsequent distancing from that politics.

7   See especially Peter C. Hodgson, *Hegel and Christian Theology: A Reading of the Lectures on the Philosophy of Religion* (Oxford: Oxford University Press, 2005); but also Hodgson's earlier edited volume *G.W.F. Hegel: Theologian of the Spirit* (London: T&T Clark, 1997).

8   Hegel saturates all of Altizer's work, but see especially *The Genesis of God: A Theological Genealogy* (Louisville: Westminster/John Knox Press, 1993) and *Godhead and the Nothing* (Albany: SUNY Press, 2003). We shall return to Altizer at the end.

9   See again Beiser, 'Introduction', pp. 1–14.

10   Charles Taylor, *Hegel* (Cambridge: Cambridge University Press, 1975).

11   See Beiser, 'Introduction', pp. 3–5, for figures who populate this side. They include the likes of Klaus Hartmann, Robert Pippin and Robert Brandom.

12   Ibid., pp. 11–14.

13   We will see how this Hegelian term 'plasticity' is itself appropriated (plastically) by Catherine Malabou below. See Chapter 6.

14   Daniel O. Dahlstrom, following a little-known early analysis by Heidegger, elaborates four notions of negativity in Hegel: 1) a purely or wholly abstracted negativity, which abstracts from a given entity; 2) the negation of such an abstracted negativity; 3) a conditioned abstract negativity (conditioned by the split into subject and object, for example); and finally 4) an unconditioned concrete negativity, as negation of the conditioned abstract negativity ('Thinking of Nothing: Heidegger's Criticism of Hegel's Conception of Negativity', in *A Companion to Hegel*, eds Stephen Houlgate and Michael Bauer (Oxford: Blackwell, 2011), pp. 524–525). Such elaboration clearly betrays the complexity and challenge in trying to reduce Hegel's concepts to tidy packages upon which everyone can agree. Heidegger's own highly personalised elaboration of Hegel's negativity and nothingness, set out more in the form of notes than in narrativised argumentation, but which surely deserves greater attention overall, can be found in 'Hegel: Die Negativität. Eine Auseinandersetzung mit Hegel aus dem Ansatz in der Negativität (1938/39, 1941)', in *Gesamtausgabe, Band 68*, ed. Ingrid Schüssler (Frankfurt am Main: Klostermann, 1993), pp. 3–61.

15   The early Judith Butler suggests that Jean Hyppolite understands negation as an ontological status or a state of lack – see *Subjects of Desire: Hegelian Reflections in Twentieth-Century France* (New York: Columbia University Press, 1987), p. 9. We will offer some correction to this below when we address Hyppolite in Chapter 4.

16   Ibid., p. 41.

17   *Faith and Knowledge*, trans. Walter Cerf and H.S. Harris (Albany: SUNY Press, 1977), p. 65. Cf. Rebecca Comay, *Mourning Sickness: Hegel and the French Revolution* (Stanford: Stanford University Press, 2011).

18   For a more condensed exposition of Hegel's understanding of becoming, especially as drawn from the *Science of Logic*, see my contribution 'Becoming', in *Resurrecting the Death of God: The Past, Present, and Future of Radical Theology*, eds Daniel Peterson and Michael Zbaraschuk (Albany: SUNY Press, 2014).

19   Angelica Nuzzo, 'How Does Nothing(ness) Move? Hegel's Challenge to Embodied Thinking', in *The Movement of Nothingness*, pp. 89–105.

20 It is the closest we come to the participial noun, here 'negating', where the verbal force is retained explicitly in the suffix. The '-ion' suffix retains the verbal force implicitly, as is often overlooked.
21 See particularly *The Science of Knowledge*, ed. and trans. Peter Heath and John Lachs (Cambridge: Cambridge University Press, 1982), where, regrettably, the German term is usually translated simply as 'act'.
22 See ibid., Part II, pp. 120ff.
23 Aristotle, *Metaphysics*, trans. Richard Hope (Ann Arbor: University of Michigan Press, 1952, 1960). Mention is made of the centrality of movement throughout, but especially with its definition in Book Delta, and then with its elaborations in Books Kappa and Lambda. In the latter Aristotle says succinctly, 'whatever produces movement or rest is a principle and a primary being' – p. 253.
24 Hegel, *The Phenomenology of Spirit*, trans. A.V. Miller (Oxford: Oxford University Press, 1977), p. 12, §22.
25 In *De Anima*, Aristotle will speak of the soul (*psychē*) as the motivator of actuality. Here he will describe two faculties within the soul that generate movement: 'There seems, then, to be these two producers of movement, either desire or intellect, if we take the imagination as a kind of thinking. For many men follow their imaginations as against their knowledge, and in the other animals, while there is neither thought nor rationality, there is imagination. Both these things, then, are productive of locomotion, and the intellect in question is that which reasons for a purpose and has to do with action and which is distinct in its end from the contemplative intellect. All desire is also purpose directed. The object of desire is the point of departure for action' – trans. Hugh Lawson-Tancred (London: Penguin, 1986), p. 214. It thus is not surprising that, in the opening of the *Philosophy of Mind*, Hegel was particularly impressed with Aristotle on this point: 'The books of Aristotle on the Soul, along with his discussion on its special aspects and states, are for this reason [a return to 'speculative' treatment] still by far the most admirable, perhaps even the sole, work of philosophical value on this topic. The main aim of philosophy of mind can only be to reintroduce unity of idea and principle into the theory of mind, and so reinterpret the less of those Aristotelian books' – *Philosophy of Mind*, trans. William Wallace (Oxford: Clarendon Press, 1971), p. 3. Thus, in the following *Zusatz* (trans. A.V. Miller), we read: 'Mind is not an inert being but, on the contrary, absolutely restless being, pure activity, the negating or ideality of every fixed category of the abstractive intellect' – p. 3. For a more detailed examination of Hegel's reading of the several conceptions of Aristotle's νοῦς (*Nous*) in *De Anima*, see Catherine Malabou, *The Future of Hegel: Plasticity, Temporality and Dialectic*, trans. Lisabeth During (Abingdon: Routledge, 2005), pp. 39–56, on which more below, Chapter 6.
26 'Introduction', in F.W.J. Schelling, *System of Transcendental Idealism (1800)*, trans. Peter Heath (Charlottesville: University Press of Virginia, 1978), p. xvii.
27 *The Science of Knowledge*, p. 97.
28 *Goethe's Faust*, trans. Walter Kaufmann (New York: Anchor Books, 1961), p. 153. The last line in the original German reads: 'Im Anfang war die Tat!' – p. 152. It is uncertain when exactly the two works, Fichte's and Goethe's, would have first appeared to each other. Goethe published an early version, *Faust, A Fragment*, in 1790, but had been working on it as early as 1772; he did not publish the more complete First Part until 1808. Fichte's *The Science of Knowledge*, or his *Wissenschaftslehre*, was originally not a book title as such, but a system developed over most of Fichte's career and through various publications and lectures, beginning its first public presentation in 1794.
29 *System of Transcendental Idealism (1800)*, p. 24.
30 Ibid., p. 25.

31   F.W.J. Schelling, *Philosophical Investigations into the Essence of Human Freedom*, trans. Jeff Love and Johannes Schmidt (Albany: SUNY Press, 2006), p. 26.
32   Jürgen Habermas, *The Philosophical Discourse of Modernity*, trans. Frederick Lawrence (Oxford: Polity Press, 1987), p. 27.
33   Nietzsche in *Die fröhliche Wissenschaft* understood Leibniz as the first modern philosopher to question the necessity of consciousness: '*Leibniz's* incomparable insight… that consciousness (*Bewusstsein*) is merely an *accidens* of the power of representation (*Vorstellung*) and *not* its necessary and essential attribute; so that what we call consciousness (*Bewusstsein*) constitutes only one state of our spiritual and psychic world (perhaps a sick state) and *by no means the whole of it*' – *The Gay Science*, ed. Bernard Williams, trans. Josefine Nauckhoff (Cambridge: Cambridge University Press, 2001), p. 217.
34   Lacan, in acknowledging a negativity in psychoanalysis (when it 'invoked destructive and, indeed, death instincts'), suggests the psychoanalysts 'were encountering that existential philosophy whose reality is so vigorously proclaimed by the contemporary philosophy of being and nothingness'. 'But', he goes on to explain, 'unfortunately that philosophy grasps negativity only within the limits of a self-sufficiency of consciousness, which, as one of its premises, links to the *méconnaissance* that constitutes the ego, the illusion of autonomy to which it entrusts itself' – in *Écrits: A Selection*, trans. Alan Sheridan (New York: Norton, 1977), p. 6.
35   *System of Transcendental Idealism (1800)*, pp. 230–231 (italics added). Hegel says something similar of Fichte's view of art: 'Through the aesthetic faculty we acknowledge a true union of the productive activity of the intelligence with the product that appears as given, of the Ego that posits itself as unlimited with the Ego that posits itself as being limited, or rather, a union of intelligence with nature, where, precisely for the sake of this possible union, nature has another side than that of being a product of intelligence' – Hegel, *The Difference Between Fichte's and Schelling's System of Philosophy*, trans. H.S. Harris and Walter Cerf (Albany: SUNY Press, 1977), pp. 152–153.
36   Jack Kaminsky, *Hegel on Art: An Interpretation of Hegel's Aesthetics* (Albany: SUNY, 1962).
37   Stephen Bungay, *Beauty and Truth: A Study of Hegel's Aesthetics* (Oxford: Oxford University Press, 1984); Beat Wyss, *Hegel's Art History and the Critique of Modernity*, trans. Caroline Dobson Saltzwedel (Cambridge: Cambridge University Press, 1999).
38   David James, *Art, Myth and Society in Hegel's Aesthetics* (London: Continuum, 2009).
39   Warren E. Steinkraus and Kenneth L. Schmitz (eds), *Art and Logic in Hegel's Philosophy* (Sussex: Harvester Press, 1980); William Maker (ed.), *Hegel and Aesthetics* (Albany: SUNY, 2000); Stephen Houlgate (ed.), *Hegel and the Arts* (Evanston: Northwestern University Press, 2007).
40   Berel Lang (ed.), *Death of Art* (New York: Haven, 1984), and Danto's essay in this volume, 'The Death of Art' (pp. 5–38).
41   Robert Wicks, *Hegel's Theory of Aesthetic Judgment* (New York: Peter Lang, 1994); Allen Speight, *Hegel, Literature and the Problem of Agency* (Cambridge: Cambridge University Press, 2001); Benjamin Rutter, *Hegel on the Modern Arts* (Cambridge: Cambridge University Press, 2010); William Desmond, *Art and the Absolute: A Study of Hegel's Aesthetics* (Albany: SUNY Press, 1986).
42   See, as but examples: Dieter Henrich, 'Art and Philosophy of Art Today: Reflections with Reference to Hegel', in *New Perspectives in German Literary Criticism: A Collection of Essays*, eds. R.E. Amacher and V. Lange (Princeton: Princeton University Press, 1979), pp. 107–33; Robert B. Pippin, 'What Was Abstract Art? (From the Point of View of Hegel)', in *Critical Inquiry* 29/1 (Autumn 2002, 575–598), and a slightly modified version in Stephen Houlgate (ed.), *Hegel and the Arts*, pp. 244–270; Stephen Houlgate,

'Hegel and the "End" of Art', *Owl of Minerva*, 29/1 (Fall 1997), pp. 1–21; and Werner Hamacher, '(The End of Art with the Mask)', in *Hegel After Derrida*, ed. Stuart Barnett (London: Routledge, 1998), pp. 105–130.

43 William Desmond, 'Art and the Absolute Revisited: The Neglect of Hegel's Aesthetics', in Maker (ed.), *Hegel and Aesthetics*, pp. 1–12 – 'Hegel's concept of art does not fit easily with post-Kantian aestheticism, of which post-Nietzschean aestheticism is one heir' – p. 2.

44 Hegel, *Aesthetics: Lectures on Fine Art, Vol. I*, trans. T.M. Knox (Oxford: Oxford University Press, 1975), p. 11.

45 Ibid., p. 7.

46 Jean-Luc Nancy, 'Why are There Several Arts and Not Just One?', in *The Muses*, trans. Peggy Kamuf (Stanford: Stanford University Press, 1996), p. 37.

47 Giorgio Agamben, *The Man Without Content*, trans. Georgia Albert (Stanford: Stanford University Press, 1996), p. 55. In more Hegelian terms: 'Artistic subjectivity without content is now the pure force of negation that everywhere and at all times affirms only itself as absolute freedom that mirrors itself in pure self-consciousness' (p. 56).

48 G.W.F. Hegel, *Elements of the Philosophy of Right*, ed. Allen W. Wood, trans. H.B. Nisbet (Cambridge: Cambridge University Press, 1991), §7 of the Introduction: '"I" determines itself in so far as it is the self-reference of negativity' – p. 41 – which owes more than a little to Fichte; or again in the *Anmerkung*: 'The task of proving and explaining in more detail this innermost insight of speculation – that is, infinity as self-referring negativity, this ultimate source of all activity, life, and consciousness – belongs to *logic* as purely speculative philosophy' – pp. 41–42.

CHAPTER 1: NEGATION'S ART IN *PHENOMENOLOGY OF SPIRIT*

1 On reading Hegel and the *Phenomenology* in light of tragic drama, see Karin de Boer, *On Hegel: The Sway of the Negative* (Houndmills: Palgrave Macmillan, 2010), pp. 17–25, e.g.

2 As most famously portrayed in M.H. Abrams' 'Hegel's "Phenomenology of Spirit": Metaphysical Structure and Narrative Plot' in *Natural Supernaturalism: Tradition and Revolution in Romantic Literature* (New York: W.W. Norton, 1971), pp. 225–237.

3 *The Phenomenology of Spirit*, p. 55, §86.

4 Frederick Beiser, *Hegel* (New York: Routledge, 2005), p. 171.

5 *Phenomenology*, p. 15, §27. 'hat est sich durch einen langen Weg hindurchzuarbeiten' – G.W.F. Hegel, *Phänomenologie des Geistes, Werke [Band] 3*, in *Werke in zwanzig Bänden*, eds E. Moldenhauer and K.M. Michel (Frankfurt: Suhrkamp, 1970), p. 31. Miller's English translation hereafter will be abbreviated *PS*; the original German *PG*.

6 *PS*, p. 488, §803.

7 Hegel's view of the novel, as he would have known it in the early nineteenth century, was, if the *Lectures on Aesthetics* are anything to go by, rather muted and uncomplimentary. However, in one of his latest writings (1828) on the author Hamann, he saw a close link between drama and the novel. In one of his notebook entries on this literary figure, Hegel wrote: '*Hamann* – drama or novel. The novel represents for us the position of the drama – since [man today] must settle [the] conflict of the characters within himself.' *Hegel on Hamann*, trans. Lisa Marie Anderson (Evanston: Northwestern University Press, 2008), p. 55.

8 *PS*, p. 1, §1.

9   Schiller, who had no small influence on Hegel, as we will see further below, employs a metaphor that is halfway to Hegel's here, by way of the most famous of Deist images: 'When the mechanic has the works of a clock to repair, he lets the wheels run down; but the living clockwork of the State must be repaired while it is in motion, and here it is a case of changing the wheels as they revolve.' *On the Aesthetic Education of Man*, trans. Reginald Snell (Mineola, NY: Dover, 2004), pp. 29–30.
10 PS, p. 2, §2.
11 For Hegel's understanding of systematic philosophy, of the relation of the *Phenomenology* as a fragment to that system, and of itself as both a fragment *and* a system, see Jon Stewart, 'Hegel's *Phenomenology* as a Systematic Fragment', in Beiser, *The Cambridge Companion to Hegel and Nineteenth Century Philosophy*, pp. 76–80.
12 PS, p. 424, §698. The phrase in German is 'Gedanken, der sich gebärendes', literally 'to give birth to itself' – PG, p. 512.
13 For general readings within Hegelian scholarship of disunity in the *Phenomenology*, see again Stewart, 'Hegel's *Phenomenology* as a Systematic Fragment', esp. pp. 80–89. See also Fredric Jameson, *The Hegel Variations: On the Phenomenology of the Spirit* (London: Verso, 2010). For Hegel's own later dissatisfaction with the text, see Michael N. Forster, *Hegel's Idea of a Phenomenology of Spirit* (Chicago: University of Chicago Press, 1998), pp. 547ff.
14 PS, p. 416, §683.
15 In the *Philosophy of Mind* (§562), Hegel will speak only of 'the close connection of art with the various religions' (p. 296). In the *Lectures on Aesthetics*, Hegel will say that the beautiful is 'spirituality given shape', and thus that the content of art and the content of religion are identical, even if their form differs. (See Chapter 3 below.) Or he will say that religion is art's 'presupposition'. But he will no longer speak explicitly of art *as* religion, or religion *as* art, or 'Spirit as *Artist*' – (PS, p. 424, §698).
16 'He should not strive for form at the expense of his reality, nor for reality at the expense of form; he should rather seek absolute being through determined being, and determined through infinite being' – *On Aesthetic Education*, p. 73 (Fourteenth Letter).
17 Ibid., p. 74. '…der Spieltrieb also würde dahin gerichtet sein, die Zeit *in der Zeit* aufzuheben, Werden mit absolutem Sein, Veränderung mit Identität zu vereinbaren.' The German used here and hereafter comes from Friedrich Schiller, 'Über die ästhetische Erziehung des Menschen, in einer Reihe von Briefen', in *Schillers sämtliche Werke, vierter Band* (Stuttgart: J.G. Cotta'sche Buchhandlung, 1879), as accessed from Projekt Gutenberg – DE, http://gutenberg.spiegel.de/buch/3355/1 (last accessed 1 May 2013).
18 Ibid.
19 Ibid., p. 101. For the later Hegel's understanding of Schiller's reconciliatory sense here, see the Introduction to *Aesthetics, Vol. I*, pp. 61–63.
20 PS, p. 412, §677.
21 We will see below in Chapter 5 how Derrida engages with these movements in his monograph on Hegel, *Glas*.
22 PS, p. 424, §698 (italics added).
23 Ibid., §699.
24 PS, p. 429, §708; '*reine Tätigkeit*' – PG, p. 517.
25 What Walter Benjamin perhaps calls 'pure language' in 'The Task of the Translator' – *Illuminations*, ed. H. Arendt, trans. H. Zohn (New York: Schocken Books, 1968).
26 William Shakespeare, 'The Winter's Tale', in *The Complete Works*, gen. ed. Alfred Harbage (New York: Viking, 1969), V.iii (p. 1367).
27 PS, p. 453, §748. For Hegel, in the 'comic consciousness… all divine being returns', that is, 'it is the complete *alienation* of *substance*' – PS, p. 455, §752.
28 PS, p. 452, §747.

29 *On Aesthetic Education*, p. 122 (Twenty-fifth Letter).
30 'The Winter's Tale', V.iii (p. 1367).
31 PS, p. 416, §683.
32 PS, p. 429, §708.
33 Together they become the 'substance-subject' that Catherine Malabou seizes upon in her work on Hegel – see below, Chapter 6.
34 PS, p. 56, §87; PG, p. 80.
35 *On Aesthetic Education*, p. 88.
36 Ibid., p. 91. The typographical emphasis on negation here is in fact Schiller's own.
37 Ibid., p. 98 (italics added).
38 Ibid., p. 99.
39 See, e.g., Introduction to *Aesthetics, Vol. I*, pp. 61–63.
40 PS, pp. 9–10, §17.
41 PS, p. 10, §18, 19.
42 PS, p. 11, §21; PG, p. 25.
43 PS, p. 12, §22; PG, p. 26.
44 PS, pp. 18–19, §32.
45 PS, p. 19, §32 – 'd. h. nur überhaupt *seiende* Unmittelbarkeit aufhebt und dadurch die wahrhafte Substanz ist, das Sein oder die Unmittelbarkeit, welche nicht die Vermittlung ausser ihr hat, sondern diese selbst ist' – PG, p. 36. The German verb translated as 'tarrying' is *verweilen* – to linger, dwell, rest upon or stay.
46 PS, p. 21, §37.
47 PG, p. 114.
48 Agamben, *The Man Without Content*, p. 67.
49 *Aesthetics, Vol. I*, p. 11.
50 Agamben, *The Man Without Content*, p. 91.
51 Jean-Luc Nancy, *Hegel: the Restlessness of the Negative*, trans. Jason Smith and Steven Miller (Minneapolis: University of Minnesota Press, 2002), p. 3.

CHAPTER 2: NEGATION'S LOGIC IN *SCIENCE OF LOGIC*

1 The text was actually published between 1812 and 1816, and became known as the 'Greater Logic' when in 1817 a shortened version, or 'outline', emerged, *The Encyclopaedia Logic* – Part 1 of a three volume *Encyclopaedia of Philosophical Sciences*, revised in 1827 and 1830 – which became known as the '*Lesser Logic*'. Our focus in this chapter will be on the '*Greater*', since it is here where the question of negation receives its full treatment.
2 English-speaking scholars of Hegel continue to vacillate between translating *Begriff* as Notion or Concept, just as they do between translating *Geist* as Spirit or as Mind.
3 Hegel's *Science of Logic*, trans. A.V. Miller (Amherst, NY: Humanity Books, 1969), p. 28. Hereafter abbreviated as *SL*.
4 SL, p. 48.
5 Charles Taylor, *Sources of the Self* (Cambridge, Mass.: Harvard University Press, 1989), p. 374.
6 Arthur Danto uses the words of Clement Greenberg, the art critic most acclamatory of abstract expressionism, to describe this kind of 'abstraction': 'In turning his attention away from the subject matter of common experience, the poet or artist turns it upon the medium of his own craft.' Danto then explicates: 'This in effect means a transformation, in the case of painting at least, from representation to object, and from content,

accordingly, to surface, or to paint itself. This, Greenberg insists, "is the genesis of the abstract", but it is a special kind of abstractness, what one might call the *material abstract*, where the physical properties of the painting – its shape, its paint, its flat surface – become the inevitable essence of the painting as art' – Arthur C. Danto, *After the End of Art: Contemporary Art and the Pale of History* (Princeton: Princeton University Press, 1997), p. 72.

7   *SL*, p. 44.
8   *SL*, p. 49.
9   Here again, the English language thwarts translation. *Wissenschaft* as Hegel understands it refers to the comprehensive knowing that attends to any one field or subject – not so much in a quantitative sense (one knows all there is to know) but more in the consummative sense (one appropriates such knowledge in the fullest sense of appropriation). Here, perhaps unexpectedly, the biblical sense of knowing is closer to the mark, not only in the sense of consummate sexual intimacy, but also in the sense of approval, guardianship and protection. For a further explication of this term, see Quentin Lauer's discussion in relation to the absolute knowing of the *Phenomenology* in *A Reading of Hegel's Phenomenology of Spirit* (New York: Fordham University Press, 1976, 1993), pp. 26–28.
10 *SL*, p. 49.
11 For Heidegger's reading of Hegel here, see his *Hegel's Phenomenology of Spirit*, trans. Parvis Emad and Kenneth Maly (Bloomington: Indiana University Press, 1988). In the concluding pages of this work, Heidegger writes: 'As we have said, we are dealing here with the explication of *a new concept of being*.' Though he links this new concept back to ancient classical philosophy (the Parmenides–Heraclitus and Plato–Aristotle pairings in particular), and points out the close relationship then with being and *eidos*, or idea, he nevertheless claims, not unlike Jesus did of the Hebrew law, that this 'new concept of being is the old and ancient concept in its most extreme and total completion'. He thus goes on to state that '*the science of the phenomenology of spirit is nothing other than the fundamental ontology of absolute ontology*, or onto-logy in general', and even that, therefore, Hegel's *Phenomenology* 'is the last stage of a possible justification of ontology' (p. 141). For more on the relationship between Hegel and Heidegger, see David Kolb, *The Critique of Pure Modernity: Hegel, Heidegger and After* (Chicago: Chicago University Press, 1986), and Karin de Boer, *Thinking in the Light of Time: Heidegger's Encounter with Hegel* (Albany: SUNY Press, 2000).
12 On this point, see Stephen Houlgate, 'Hegel's Logic', in Beiser, *The Cambridge Companion to Hegel and Nineteenth Century Philosophy*, p. 121.
13 *SL*, p. 49.
14 William Desmond is among the few Hegelian philosophers working in English to grasp and appreciate this formative understanding of Hegelian logic and its *Begriff* in relation to art. In his *Art and the Absolute*, he describes the concreteness of art in four formative ways: 'it implicates the idea of dynamic origination'; 'it makes concrete a certain process of emergence'; 'it articulates itself into a shape of a certain embodied formation'; and 'it brings to light a rich wholeness and concrete universality' (p. 23). He then applies these to philosophical concepts: 'Philosophical concepts must grasp this process of emergence, for this process is the active formation of intelligibility' (p. 24). He then summarises: 'an aesthetic dimension infiltrates the Hegelian style of logical thought. It is not just that a philosophical system lies at the back of his aesthetics, though this is true. It is that Hegel's notion of philosophical system, particularly in its emphasis upon dynamic form, organic unity and wholeness, has an unmistakably aesthetic ring' (p. 30). So he can conclude: 'In presenting the developing drama of essential thought forms, the *Logic* reveals, as it were, the philosophical art of reason itself' (p. 31). See the entirety of his Chapter 2 for the working out of this 'art'.

15 *SL*, p. 63.
16 See *SL*, pp. 61–63.
17 On the question of cognition as mere instrument, cf. the opening of the *Phenomenology*, Introduction, pp. 46–48.
18 Or as the pure *act* of objectivity. Hegel sees Kant's successors (Fichte et al.) as at least trying to bring focus upon this act as activity, a move in the right direction. See *SL*, p. 62, and Hegel's own footnote (1) on the expression 'objectifying act of the ego' on the same page. Unfortunately, it does not yet, for Hegel, go far enough.
19 Houlgate, 'Hegel's Logic', pp. 118–119.
20 The Latin root here is *genero* (to beget, procreate, engender, produce) and/or its cognate *gigno* (to produce, give birth to, bear, beget, bring forth), which declines by means of the principle part '*gen-*'. The Latin itself derives from the Greek verb *gennao* or the Greek noun *genesis* of the same meanings.
21 This term, in itself, would retain, simultaneously, both the nominal and adjectival force of its grammatical functions, even while it signifies the operations of a genitive function. But it also collapses the signifier/signified distinction: its grammatical function as a signifier (noun and/or adjective) is precisely what it signifies – the conversion of the genitive proper into a noun/adjective. So too its semantic function: the generative nature of this conversion works both sides of the signifier/signified divide. In fact, as we will see below, it is the very *genesis* of this divide as unified ('co-').
22 The 'cogenitive' we might see as an internal feature of dialectic, but not identical or co-extensive with it. The dialectic sublates, that is, creates a *higher* term. The cogenitive, on the other hand, makes mutual, that is, creates the propositional grounds for an equivalence. In dialectical terms, it is not that the coming together of the thesis and antithesis creates a synthesis; rather, the thesis becomes the antithesis and the antithesis the thesis. Grammatically, one would write: 'the thesis of antithesis' and vice versa. That the cogenitive, even as an internal operation, threatens to *undermine* the dialectic, by not letting the sublation have the fullness of its higher valence, is something we need to hold open for consideration. For it would go to the heart of negation.
23 *SL*, p. 63.
24 As stated as early as the *Phenomenology*'s Preface (p. 15, §27), as we have already seen above (Chapter 1, p. 24 and note 5).
25 *SL*, p. 844. The circularity is also evident with *The Science of Logic* itself: we noted earlier that in the Preface to the First Edition, where Hegel describes the natural succession from the *Phenomenology* to the *Logic*, 'Consciousness is spirit as a concrete knowing [*Wissen*]', a knowing which is also a 'pure knowing [*reinen Wissen*]', whose essentialities are in turn 'pure thoughts [*reinen Gedanken*], spirit thinking its own essential nature' through, in the end, logic – *SL*, p. 28; *Wissenschaft Der Logik I, Werke [Band] 5*, in *Werke in zwanzig Bänden*, eds E. Moldenhauer and K.M. Michel (Frankfurt: Suhrkamp, 1969), p. 17. Hereafter abbreviated *WL-I*. See also Desmond, *Art and the Absolute*, p. 28.
26 We find this paradox captured in the following remarkable passage of the *Science of Logic*'s opening section of Book One ('With What Must the Science Begin?'), all of which should really be quoted: 'It must be admitted that it is an important consideration – one which will be found in more detail in the logic itself – that the advance is *a retreat into the ground*, to what is *primary* and *true*, on which depends and, in fact, from which originates, that with which the beginning is made. Thus consciousness on its onward path from immediacy with which it began is led back to the absolute knowledge of its innermost *truth*. This last, the ground, is then also that from which the first proceeds, that which at first appeared as immediacy... The essential requirement for a science of logic is not so much that the beginning be a pure immediacy, but rather that the whole of the science be within itself a circle in which the first is also the last and the last is

also the first' – p. 71. One advances to the beginning. Anew. And 'the line of scientific advance becomes a *circle*' – p. 72. For more on circles in Hegel's *oeuvre*, see Michael Inwood, *Hegel* (London: Routledge, 1983, 2002), pp. 149–154, 317–320.
27 But it is not the first. As far back as the *Wissen und Glauben* of 1802, Hegel in the Introduction had written, in relation to absolute finitude and absolute infinity: 'there is no reflection on their opposition, which is to say that the opposition is not objective: the empirical is not posited as the negativity for the concept nor the concept as negativity for the empirical nor the concept as that which is in itself negative. When abstraction achieves perfection there is reflection on this opposition, the ideal opposition becomes the objective, and each of the opposites is posited as something which is not what the other is. Unity and the manifold now confront one another' – *Faith and Knowledge*, p. 62.
28 *SL*, p. 68.
29 *SL*, p. 67, italics added.
30 *SL*, pp. 68–69. 'Further, the *progress* from that which forms the beginning is to be regarded as only a further determination of it [the beginning]' – p. 71.
31 Desmond: 'Hegel almost obsessively dwells on the question of origin of philosophy and repeatedly returns to the question of beginnings, both in the *Phenomenology* and the *Logic*' – *Art and the Absolute*, p. 24.
32 Aristotle, *Metaphysics*, p. 87, Book Delta, 1013a.
33 *Metaphysics*, p. 10, Book Alpha, 983b.
34 See for example G.S. Kirk, J.E. Raven and M. Schofield (eds), *The Presocratic Philosophers*, 2nd Ed. (Cambridge: Cambridge University Press, 1957, 1983), pp. 90–91, 106–109.
35 *hothen hé arché tés kinéseos* – Aristotle, *Metaphysics*, Book Alpha, 983a, 984a, 988a, 994a, 1064a.
36 *Metaphysics*, p. 16, Book Alpha, 986b.
37 Philosophy as the question of the first principle: 'But what we intend to say in the present account is that all men take what is called "wisdom" [*gnosis*] to be concerned with the first principles [*arché*] of explanation [*aitia*]' – *Metaphysics*, p. 5, Book Alpha, 981b.
38 *SL*, p. 69.
39 Ibid. The *Encyclopaedia Logic* deals with the unity of immediacy and mediation more extensively, as it arises first in relation to critical philosophy (Kant and Jacobi). See its §61–78. In fact, in the Remark of §65 Hegel makes the bold statement: 'The entire second part of the *Logic*, the doctrine of *Essence*, deals with the essential self-positing unity of immediacy and mediation' – trans. T.F. Geraets et al. (Indianapolis: Hackett, 1991), p. 115.
40 *SL*, p. 69.
41 'The beginning is *logical* in that it is to be made in the element of thought that is free and for itself, in *pure knowing*. It is *mediated* because pure knowing is the ultimate, absolute truth of *consciousness*' – *SL*, p. 68.
42 *SL*, pp. 68, 70. In the latter reference, the phrase is put in apposition to 'what is only directly involved in the simplest of all things [*Allereinfachsten*]'.
43 *SL*, p. 70. Cf. 'it lies in the *very nature of a beginning* that it must be being and nothing else' – p. 72. Here, at this beginning, might lie one further permutation of the cogenitive: 'beginning itself'. For beginning here is not only beginning *as such*, beginning *qua* beginning, beginning *in itself*, but also the start of *itself*, a beginning *for itself*. Thus 'beginning itself' must be read as beginning *in-and-for itself*, where the itself is nothing other than beginning, in every sense. And this beginning, this itself, is, and can only be, pure being.
44 The German *stimmen* can also mean 'to tune', related to *Stimme*, 'voice'; one tunes the voice of the instrument, beginning with the voice itself. This alternate meaning,

which is still cognate to the idea of certainty, Hegel may have had in mind when he later quotes Jacobi's words: 'Whence does the vowel get its consonant, or rather how does its *soundless*, uninterrupted *sounding* interrupt itself and break off in order to gain at least a kind of "self-sound" (vowel), an *accent?'* – *SL*, pp. 95–96.

45 *SL*, p. 129.
46 *SL*, p. 83.
47 *SL*, p. 90.
48 See Maurice Blanchot, *The Instant of My Death*, and Jacques Derrida's response to it, *Demeure: Fiction and Testimony*, both trans. Elizabeth Rottenberg (Stanford: Stanford University Press, 2000).
49 Hegel makes clear that this sublation is not to be seen as 'reciprocal': 'the one does not sublate the other externally – but each sublates itself in itself and is in its own self the opposite of itself' – *SL*, p. 106.
50 *SL*, p. 106.
51 *SL*, p. 106. Cf. *The Encyclopaedia Logic*: 'As their unity, *becoming* is the true expression of the result of being and nothing; it is not just the *unity* of being and nothing, but it is inward *unrest* – a unity which in its self-relation is not simply motionless, but which, in virtue of the diversity of being and nothing which it contains, is inwardly turned against itself' – Fourth Remark, p. 143, §88.
52 Or as Hegel says later, 'in the negative as such there lies the ground of becoming, of the unrest of self-movement' – *SL*, p. 166.
53 *SL*, p. 107. Translation slightly altered: 'die *Bestimmtheit, aus der es herkommt, noch an sich*' – *WL-I*, p. 114.
54 Heidegger's interest is predominantly in *Nichts*, rather than *Negation*, it should be noted. See especially 'What is Metaphysics?', trans. David Farrell Krell, in *Pathmarks*, ed. William McNeill (Cambridge: Cambridge University Press, 1998), pp. 82–96. For a more recent discussion of Heidegger's Nothing in relation to Hegel, see Roberto Morani, 'Heidegger, Hegel und die Frage des Nichts', in *Nichts – Negation – Nihilismus: Die europäische Moderne als Erkenntnis und Erfahrung des Nichts*, eds Allessandro Bertinetto and Christoph Binkelmann (Frankfurt am Main: Peter Lang, 2010), pp. 79–92. See also above, Introduction, note 14.
55 *SL*, p. 113. Hegel will claim this definition for Spinoza's '*determinatio negatio est*' ('determination is negation' (Epistola, 50, *Opera*, IV), which he will alter to '*Omnis determinatio est negatio*'. Of course, Spinoza had only in mind the question of an infinite substance, whose very limitlessness precludes determination of any kind, for this would introduce finitude. He did not have in mind Hegel's sense of nothing as constitutive of being. For more on the Hegel's reading of Spinoza here, see Yitzhak Y. Melamed, '"Omnis determinatio est negatio": determination, negation, and self-negation in Spinoza, Kant, and Hegel' in Eckart Förster and Yitzhak Y. Melamed, eds *Spinoza and German Idealism* (Cambridge: Cambridge University Press, 2012).
56 See *SL*, pp. 114–116.
57 Karin de Boer, in her *On Hegel: The Sway of the Negative*, will speak, alongside others, of the 'negation of negation' in terms of an 'absolute negativity', which she convincingly sources in an early, lesser-known work of Hegel's, *Essay on Natural Law* (see her Introduction and Chapter 3 esp.). But an emphasis on the restive nature of beginning in nothing, and subsequently in negation, would challenge any notion of 'absolute' here.
58 There are several worthy commentaries on *The Science of Logic*, but see in particular Stephen Houlgate's *The Opening of Hegel's Logic: From Being to Infinity* (West Lafayette, IN: Purdue University Press, 2006), which itself is limited only to the first two chapters of Book One, 'Doctrine of Being', ending with the concept of 'infinity'.

59 For example, *SL*, pp. 69, 81.
60 *SL*, p. 385; *WL-I*, p. 457.
61 Charles Taylor remains strong on this point: see *Hegel*, pp. 256–257 esp.
62 *SL*, p. 399.
63 *SL*, p. 400.
64 *SL*, pp. 417–418.
65 *SL*, p. 439.
66 Such can be found, for reflection, in the examples of Rodolphe Gasché, *The Tain of the Mirror: Derrida and the Philosophy of Reflection* (Cambridge, Mass.: Harvard University Press, 1986), pp. 13–65 esp.; and of Stephen Houlgate, 'Essence, Reflexion and Immediacy in Hegel's *Science of Logic*', in *A Companion to Hegel*, eds Stephen Houlgate and Michael Bauer (Oxford: Blackwell, 2011), pp. 137–158. For the case of contradiction, see for example Sonsuk Susan Hahn, *Contradiction in Motion: Hegel's Organic Concept of Life and Value* (Ithaca: Cornell University Press, 2007).
67 Immanuel Kant, *Critique of Pure Reason*, trans. Norman Kemp Smith (Houndmills: Macmillan, 1933), pp. 50–52, A10, B14). For a closer analysis of the hidden and disruptive nature of the power at work in the *a priori* synthetic judgements, see my *Poetics of Critique: The Interdisciplinarity of Textuality* (Aldershot: Ashgate, 2003), pp. 43–62.

CHAPTER 3: ART'S NEGATION IN AESTHETICS

1 See *The Philosophy of History*, trans. J. Sibree (New York: Dover, 1956), pp. 250–274.
2 Hegel did not apply the cunning of reason only to history, where individual human passions are converted to the overall purpose of world-historical *Geist* (see ibid., p. 33); he also applied it to logic, and particularly to teleological aspects in logic. See e.g. *SL*, p. 746. In a Remark of the *Lesser Logic*, Hegel even goes so far as to claim that divine Providence behaves with 'absolute cunning' in this respect: 'God let men, who have their particular passions and interests, do as they please, and what results is the accomplishment of *his* intentions, which are something other than those whom he employs were directly concerned about' – *Encyclopaedia Logic*, p. 284, §209.
3 We would need to be careful here. Hegel claims that such cunning is a 'mediating activity' – *Encyclopaedia Logic*, p. 284, §209. But as we saw in the last chapter, the logic of negation is that its mediation is also an immediacy, the immediacy of a beginning. If negation operates its 'purpose' without being present (its 'end' is always contradictory: negation of itself, as beginning), it is because its purposive activity, generation, is always in the process of becoming. This would be the paragon of Hegelian cunning: to be there and not there at the same time.
4 H.G. Hotho was the student in question. See Robert Pippin, 'The Absence of Aesthetics in Hegel's Aesthetics', in Beiser, *The Cambridge Companion to Hegel and Nineteenth Century Philosophy*, fn 3, for a brief discussion of the controversy surrounding the credibility of Hotho's compilation. There were other earlier lectures on art that Hegel had given, in Heidelberg (1817–18) and in Berlin (1820–21). The first three of the Berlin lectures have now been published separately in German, though they await translation into English.
5 *Philosophy of Right*, p. 380, §360.
6 As we will see at the end of our discussion, there is debate around who wrote a fragment entitled 'The Oldest *Systemprogramm* of German Idealism', written in 1797, in which a call for a new 'aesthetic philosophy', as a 'new mythology', is issued. Whatever the truth about authorship, the fragment is largely a response to Kant's *Critique of Practical Reason* (and not to the *Critique of Judgment*), and does not treat the question

of art on its own terms. An earlier possible exception to Hegel's general disregard for 'aesthetics' is a short essay he had written when still a teenager in 1788, entitled 'On a Few Characteristic Differences Between Ancient and Modern Poets' – 'Über einige charakteristische Unterschiede der alten Dichter [und der neueren]', in *Dokumente zu Hegels Entwicklung*, ed. J. Hoffmeister, 2nd ed. (Stuttgart/Bad Cannstatt: Frommann-Holzboog, 1974), pp. 48–51.

7   *Philosophy of Mind*, p. 293, §556.
8   Ibid., p. 297, §§562, 563. The ideas are difficult here, and rely on unstated assumptions about the relationship between beauty and freedom, as informed particularly by Kant and his *Critique of Judgment*. What Hegel is essentially saying is that the *form* of beauty is not possible to grasp except by way of an internal freedom of spirit (Kant would link this with judgement). If form and content are to merge in the absolute self-actualisation of Spirit, as the earlier *Logic* told us they must, they can only do so by way of such internal freedom released by the beautiful shape. The devotion to, say, bones (to take the 'hideous' example Hegel gives us) keeps the devotee bound to the crass sensuousness of materiality, which can only remain unspiritual and unabsolute. Beauty, by contrast, allows Spirit to become conscious of its form, as beauty.
9   Ibid., §563.
10 Martin Gammon, 'Modernity and the Crisis of Aesthetic Representation in Hegel's Early Writings', in Maker (ed.), *Hegel and Aesthetics*, p. 148.
11 Since the word 'theory' has its etymological roots (the Greek θεωρία (*Theoria*)) in spectatorship as well – the verbal form meant 'to look at', 'behold', 'take in', as one does a theatrical performance: in ancient Athens, drama was a place for *theoria*, either actively (audience) or passively (performer) – we might then say that Hegel was always trying to rewrite the notion of theory too: it can never remain in the abstract, but, like art, must engage the concrete as itself. Hegel, in his art lectures, had said that drama was 'the most perfect totality of content and form' (*Aesthetics, Vol. II*, p. 1158) because what was being beheld (on stage) was in fact the inner act of beholding (in consciousness of Spirit). Likewise, in the concluding section on the 'theorem' in the *Science of Logic*, Hegel had written: 'The Idea, in so far as the Notion [*Begriff*] is now *explicitly* determined in and for itself, is the *practical* Idea, or *action*' (p. 818). These two traditional rivals, theory and practice, come together no more explicitly than in a telling passage in the *Phenomenology* in the section of 'Religion in the Form of Art': 'The common element in the work of art, viz. that it is produced in consciousness and is made by human hands, is the moment of the Notion [*Begriff*] existing *qua* Notion, which stands in contrast to the work. And if this Notion, whether in the shape of artist or spectator, is unselfish enough to declare the work of art to be in its own self absolutely inspired, and to forget himself as performer or as spectator, then against this we must stick to the Notion of Spirit which cannot dispense with the moment of being conscious of itself. This moment, however, stands in contrast to the work because in this initial duality of itself Spirit gives the two sides their abstract, contrasted characters of *action* and of being a Thing, and their return into the unity from which they proceeded has not yet come about' – p. 429, §708.
12 Schiller, *On Aesthetic Education*, p. 99.
13 *Aesthetics, Vol. I*, p. 1 – hereafter abbreviated *A-I*.
14 Pippin, 'The Absence of Aesthetics', pp. 395–396. Ironically, Pippin's essay in this volume comes directly after Allen Speight's essay ('Hegel's Aesthetics: The Practice and "Pastness" of Art') in which the exact opposite is argued: Hegel gives to art a new autonomy by connecting to a Spirit that can remain independent in its engagement with another (p. 380), like, we suppose, the cunning of reason (and so now, perhaps, the cunning of art). Together the two essays neatly show the dividing line in thinking about

Hegel's relation to art: either it is part of the system as *aesthetics*, and therefore should be treated *aesthetically* (Speight, Bungay, e.g.); or it is part of the philosophical system as a unified conceptual whole, and should be treated *organically* (Pippin, Desmond, e.g.).

15 *Philosophy of Mind*, p. 294, §558.
16 *A-I*, p. 55.
17 Desmond, *Art and the Absolute*, p. 12. Desmond sees in this a possible new form of imitation: art is really only an imitation of 'the conceptual original constituted in proper form in philosophy' (ibid.). He sees both a positive and negative side to this new form: art can lead towards, but also still conceal, the philosophical form.
18 *A-I*, p. 8; but also p. 38, 168, etc.
19 *A-I*, p. 38.
20 *A-I*, pp. 12–13; *Vorlesungen über die Ästhetik I, Werke [Band] 13*, in *Werke in zwanzig Bänden*, eds E. Moldenhauer and K.M. Michel (Frankfurt: Suhrkamp, 1970), pp. 27–28. Hereafter abbreviated as *VA-I*.
21 In Volume Two of the *Science of Logic*, 'The Doctrine of the *Begriff*', Hegel writes, now with an added significance in our present context of sensuous artwork, of *Begriff*'s incompletion in abstract thinking, and its necessary completion in 'the sublating and reduction of that [sensuous] material as *mere phenomenal appearance* to the *essential*, which is manifested only in the Notion [*Begriff*]' (p. 588). He furthers the thought: 'the Notion in its formal abstraction reveals itself incomplete and through its own immanent dialectic passes over into reality; but it does not fall back again onto a ready-made reality confronting it and take refuge in something which has shown itself to be the unessential element of Appearance because, having looked around for something better, it has failed to find it; on the contrary, *it produces the reality from its own resources*' (pp. 591–592, italics added).
22 *A-I*, p. 58.
23 *A-I*, p. 152.
24 For example, Desmond, *Art and the Absolute*, pp. 103–165; Edward Halper, 'The Logic of Art: Beauty and Nature', in Maker, *Hegel and Aesthetics*, pp. 187–202; Bungay, *Beauty and Truth*.
25 *On Aesthetic Education*, p. 73 (Fourteenth Letter).
26 Ibid., p. 87.
27 Kant says: 'The universal communicability of a pleasure involves in its very concept that the pleasure is not one of enjoyment arising out of mere sensation, but must be one of reflection. Hence, aesthetic art, as art which is beautiful, is one having for its standard the reflective judgement and not organic sensation' – *Critique of Judgment*, trans. James Creed Meredith (Oxford: Clarendon Press, 1952), p. 166.
28 *On Aesthetic Education*, p. 88. Italics added.
29 Ibid.
30 Ibid., p. 89.
31 Ibid., p. 91.
32 Ibid., p. 92. This contradiction can be seen also in light of Fichte's conception of *Tathandlung* (above, Introduction), where the 'fact/act' refers to the self-production of the *Geist* as thinking mind, now through negation. This contradiction can also be seen as one of the early articulations of what has now become the classic hermeneutical circle: 'Thus we arrive, to be sure, at the whole only through the part, at the unlimited only through limitation; but we also arrive at the part only through the whole, at limitation only through the unlimited' (ibid.).
33 Ibid., p. 93.
34 Ibid., p. 98.

35 Ibid.
36 Ibid., pp. 98–99. Italics added.
37 This 'extrapolation' is not always as consistent as we might expect from Hegel – suggesting that the redaction of the lecture notes is, at best, a secondary substitute to how Hegel would have handled the material himself in book form.
38 *Phenomenology*, pp. 483–491, §795–805.
39 *SL*, pp. 756–758. Cf. *A-I*, p. 106: 'Now the Idea as such is nothing but the Concept, the real existence of the Concept, and the unity of the two.'
40 *SL*, p. 759.
41 *A-I*, p. 72. *VA-I*, p. 103.
42 *A-I*, p. 92.
43 Ibid., p. 94. By the end of this opening section, 'Position of Art in Relation to the Finite World and to Religion and Philosophy', art will 'stand on one and the same ground' as religion and philosophy (p. 101), and even if it passes over into higher forms of consciousness in religion and, ultimately, philosophy (pp. 102–103), nevertheless the content of these three forms is the same, and their realms are identical: absolute *Geist*.
44 Ibid., pp. 92–93.
45 Ibid., p. 94.
46 Ibid.
47 Ibid., p. 109.
48 Ibid., p. 97; *VA-I*, p. 134. A slightly later passage in the chapter 'The Beauty of Nature' recapitulates the same idea: 'Yet whoever claims that nothing exists which carries in itself a contradiction in the form of an identity of opposites is at the same time requiring that nothing living shall exist. For the power of life, and still more the might of the spirit, consists precisely in positing contradiction in itself, enduring it, and overcoming it. This positing and resolving of the contradiction between the ideal unity and the real separatedness of the members constitutes the constant process of life, and life *is* only by being a *process*' – *A-I*, p. 120; *VA-I*, p. 162.
49 For a discussion on an interpretation of the master–slave passage in relation to art, see my 'Artist Bound: The Enslavement of Art to the Hegelian Other', in *Literature and Theology* 25/4 (December 2011), pp. 379–392. For two more recent discussions about the end of art debate/dilemma in Hegel, see two contributions in Stephen Houlgate's *Hegel and the Arts*, Martin Donougho's 'Art and History; Hegel on the End, the Beginning, and the Future of Art', pp. 179–215, and J.M. Bernstein's 'Freedom from Nature? Post-Hegelian Reflections on the End(s) of Art', pp. 216–243. See also above, Introduction, note 42.
50 *A-I*, p. 11. See above, p. 17.
51 *A-I*, p. 101.
52 Ibid.
53 Ibid., p. 102.
54 Ibid., p. 103.
55 Ibid.

CHAPTER 4: THE RETURNING OF HEGEL AND NEGATION: SARTRE AND HYPPOLITE

1 Theodor Adorno, *Negative Dialectics*, trans. E.B. Ashton (New York: Continuum, 2007), p. 362.

2   For a fuller discussion of the ascendancy of negation in the modern West, see my *Auden's O: The Loss of One's Sovereignty in the Making of Nothing* (Albany: SUNY Press, 2013).
3   Jean-Paul Sartre, *Being and Nothingness*, trans. Hazel E. Barnes (London: Routledge, 2003), p. 35.
4   See Judith Butler, *Subjects of Desire*, particularly pp. 92ff. Sartre had attended Kojève's lectures on Hegel as a student. A more direct engagement with Hegel will occur later, and surface particularly in his *Critique of Dialectical Reason* (1960).
5   *Being and Nothingness*, p. 21.
6   Ibid., p. 619.
7   Ibid., p. 22. See also p. 206: 'We must understand that this negation – seen from the point of view of the "this" – is wholly ideal. It adds nothing to being and subtracts nothing from it. The being confronted as "this" *is* what it is and does not cease being what it is; it does not become.'
8   PS, p. 21, §37. See also above, p. 34.
9   See for example ibid., p. 69, §114.
10 For more on the question of 'why being and nothing "move"' in the *Science of Logic*'s opening, see Houlgate, *The Opening of Hegel's Logic*, pp. 272–274.
11 *Being and Nothingness*, p. 22.
12 Ibid., p. 38.
13 Ibid., p. 39.
14 Ibid.
15 Ibid., p. 40.
16 Ibid., p. 85.
17 In the Conclusion, Sartre writes of the for-itself: 'Its reality is purely *interrogative*. If it can post questions this is because it is itself always *in question*; its being is never *given* but *interrogated* since it is always separated from itself by the nothingness of otherness' – ibid., p. 639.
18 Ibid., p. 206.
19 See again *Auden's O*, Chapter 6, 'Before the Postmodern: Sartre'.
20 *Being and Nothingness*, p. 48.
21 Foucault in a 1966 interview with C. Bonnefoy, as quoted by Thomas Flynn in *Sartre, Foucault, and Historical Reason, Volume One: Toward an Existentialist Theory of History* (Chicago: University of Chicago Press, 1997), p. 237. The context of the quotation is in fact Sartre's later book on history and Marxism: '*The Critique of Dialectical Reason* is the magnificent and pathetic attempt by a man of the nineteenth century to think the twentieth century. In this sense, Sartre is the last Hegelian and, I would even say, the last Marxist.'
22 *Being and Nothingness*, p. 640.
23 We can also see this allegiance in the way Sartre construes the question, for 'founding' is not in a strict sense identical to 'generating'. But a much fuller discussion is warranted here than we can afford at present.
24 Ibid., pp. 643–644.
25 Jean Hyppolite, *Logic and Existence*, trans. Leonard Lawlor and Amit Sen (Albany: SUNY Press, 1997), p. 113.
26 *Being and Nothingness*, p. 643.
27 Ibid., p. 644.
28 Jacques Derrida, 'The Ends of Man' in *Margins of Philosophy*, trans. Alan Bass (London: Harvester Wheatsheaf, 1982), p. 116.
29 See Sartre's essay of 1947 'What is Literature', in *What is Literature? And Other Essays*, trans. Bernard Frechtman et al. (Cambridge, Mass.: Harvard University Press, 1988).

Here Sartre will say it is not just the work and its author but also the reader who is involved in the freedom that art embodies (p. 54, e.g.).
30 Butler, *Subjects of Desire*, p. 97.
31 *Logic and Existence*, p. 36.
32 Ibid., p. 87.
33 Ibid., p. 106.
34 Ibid., p. 108.
35 *Phenomenology*, p. 19, §32. (Hyppolite, *Logic and Existence*, p. 107). See also above, p. 33.
36 *Logic and Existence*, p. 113.
37 Plato, *Sophist*, 256d, as quoted in *Logic and Existence*, p. 112.
38 Plato, *Sophist*, 253b, as quoted in *Logic and Existence*, p. 113.
39 Hyppolite could have noted here that Hegel, in his section on contradiction in the *Science of Logic*, had himself addressed such a Platonic notion of alterity as 'relation' or 'relationship' in respect to contradiction. 'Ordinary thinking', as Hegel calls it, can see that between two related things or determinate realities there is a difference one to the other (a father is not a son, and a son is not a father, etc.), but it tends to pass over or forget their *negative* unity. 'It holds these two determinations over against one another and has in mind *only them*, but not their *transition*, which is the essential point and which contains the contradiction' – *SL*, p. 441.
40 *SL*, p. 439. See also above, p. 54. Hyppolite calls this 'speculative contradiction' in an earlier chapter. See, e.g., *Logic and Existence*, pp. 91–92.
41 See for example *Genesis and Structure of Hegel's Phenomenology of Spirit*, trans. Samuel Cherniak and John Heckman (Evanston: Northwestern University Press, 1974), pp. 16, 149, 167 and 421. In this text, Hyppolite quotes more than once this passage from Hegel's Preface to the *Phenomenology*: 'The realized purpose, or existent actuality, is movement and unfolded becoming; but it is just this unrest [*Unruhe*] that is the Self' – *PS*, p. 12; *PG*, p. 26.
42 Michel Foucault, 'The Discourse on Language', in *The Archaeology of Knowledge*, trans. A.M. Sheridan Smith (New York: Pantheon, 1972), p. 235. Also quoted by Stuart Barnett's 'Introduction' to *Hegel after Derrida*, ed. Stuart Barnett (London: Routledge, 1998), p. 3.

CHAPTER 5: THE TOLLING OF HEGEL AND NEGATION: DERRIDA

1 Jacques Derrida, 'Structure, Sign, and Play in the Discourse of the Human Sciences', in *Writing and Difference*, trans. Alan Bass (London: Routledge, 1978), p. 288.
2 Jacques Derrida, *Positions*, trans. Alan Bass (Chicago: University of Chicago Press, 1981; London: Athlone Press, 1987), p. 77.
3 Ibid., p. 24.
4 Jacques Derrida, 'Tympan', in *Margins of Philosophy*, p. xi.
5 Jacques Derrida, 'The Pit and the Pyramid: Introduction to Hegel's Semiology', in *Margins of Philosophy*, p. 71.
6 This sense of mediation might be extended to all of Derrida's philosophical interlocutors. Derrida is always reading another through the lens of someone else. This is, of course, a deliberate, highly calculated strategy: the 'lens' – language, discourse, writing, text, script, sign, grammatology and so on – is never not operating in the structures and activity of our thinking. Even, we are told, *prior* to the letter of the text (as much as to the letter of the law). Of course, much has been made of this midrashism. Derrida

himself sums it up in 'Edmund Jabès and the Question of the Book': 'In the beginning is hermeneutics' (*Writing and Difference*, p. 67).
7   See for example Stuart Barnett's volume *Hegel After Derrida*.
8   *Writing and Difference*, pp. 251–277.
9   Ibid., p. 252.
10 Cf. Heinz Kimmerle, 'On Derrida's Hegel Interpretation', in *Hegel After Derrida*, pp. 228–232.
11 *Writing and Difference*, p. 253.
12 Ibid., p. 255.
13 Ibid., p. 264.
14 Ibid., pp. 263 and 273 respectively.
15 Ibid., p. 257.
16 Ibid., p. 259.
17 Ibid.
18 Ibid. The original context in which Bataille said these words – 'he did not know to what extent he was right' – can be found in 'Hegel, Death and Sacrifice', *The Bataille Reader*, eds Fred Botting and Scott Wilson (Oxford: Blackwell, 1997), pp. 288–289.
19 In *Introduction to the Reading of Hegel*, Kojève writes: 'And Man is essentially *Negativity*, for Time is *Becoming*, that is, the *annihilation* of Being or Space. Therefore Man is a Nothingness that nihilates and that preserves itself in (spatial) Being only by *negating* being, this Negation being Action. Now, if man is Negativity – that is, Time – he is not eternal. He is born and he dies as Man. He is "*das Negative seiner selbst*," Hegel says. And we know what he means: Man overcomes himself as Action (or *Selbst*) by ceasing to *oppose* himself in the World, after creating in it the universal and homogeneous State; or to put it otherwise; on the cognitive level: Man overcomes himself as *Error* (or "Subject" *opposed* to the Object) after creating the Truth of "Science"' – ed. Allan Bloom, trans. James H. Nichols, Jr. (New York: Cornell University Press, 1969), p. 160.
20 See again my article 'Artist Bound: The Enslavement of Art to the Hegelian Other' for a rethinking of the artist as master.
21 'From Restricted to General Economy', p. 261.
22 Ibid.
23 The others were *Of Grammatology* (1969), *Dissemination* (1972), *The Truth in Painting* (1978), *Spurs: Nietzsche's Style* (1978) and *The Post Card: from Socrates to Freud and Beyond* (1980).
24 'Otobiographies: The Teaching of Nietzsche and the Politics of the Proper Name', in *The Ear of the Other: Otobiography, Transference, Translation*, ed. Christie McDonald, trans. Peggy Kamuf and Avital Ronell (Lincoln: University of Nebraska Press, 1988), p. 4. Here the limit operates between 'work' and 'life' or between the written corpus and the living body.
25 'This borderline – I call it *dynamis* because of its force, its power, as well as its virtual and mobile potency – is neither active nor passive, neither outside nor inside' – *Otobiographies*, p. 5.
26 'Tympan', p. xxiii.
27 'The Pit and the Pyramid', p. 92.
28 Ibid., pp. 92–93. The actual quotation in Knox's translation that is used here extends to p. 891. The original is found in *VA-III*, *Werke [Band]* 15, pp. 134–135.
29 'The Pit and the Pyramid', p. 93.
30 Cf. Alan Roughley, *Reading Derrida Reading Joyce* (Gainsville: University Press of Florida, 1999), esp. Chap. 5, 'Examples and Counter-examples: *Finnegans Wake* and *Glas*'; and Peter Mahon, *Imagining Joyce and Derrida: Between Finnegans Wake and Glas* (Toronto: University of Toronto Press, 2007).

31 One should consult John P. Leavey, Jr.'s *Glassary* (Lincoln: University of Nebraska Press, 1986), which among other things provides the much-demanded references to a text teeming with unspecified voices and quotations. See also Simon Critchley, 'A Commentary Upon Derrida's Reading of Hegel in *Glas*', in *Hegel After Derrida*, pp. 197–226, as well as several other less summary essays on the text in this volume.

32 I am referring to the English edition here, translated by John P. Leavey, Jr. and Richard Rand, and published by the University of Nebraska Press, whose page-width is well beyond the standard.

33 *Glas*, trans. John P. Leavey, Jr. and Richard Rand (Lincoln: University of Nebraska Press, 1986), p. 232. References to the text will hereafter be G. It is customary within Derridean scholarship to differentiate somehow with additional abbreviations between the two columns of each page, and sometimes even within each column. Here, following Critchley and others, we will use the letter 'a' after the page number to refer to the left-hand column, and 'b' after the page number for the right-hand column. Hence, in this case, p. 232a (even if in this instance the left-hand column is further split into two columns). The full sentence (from Kierkegaard's *Concluding Unscientific Postscript to the Philosophical Fragments*) is worth quoting: '"Hegel is also supposed to have died with the words upon his lips, that there was only one man who had understood him, and he had misunderstood him... Hegel's statement reveals at once the defect of a direct form, and hence is quite inadequate as an expression for such a misunderstanding, giving sufficient evidence that Hegel has not existed artistically in the elusive form of a double reflection."'

34 Henry Sussman, 'Hegel, *Glas*, and the Broader Modernity', in *Hegel After Derrida*, p. 260.

35 In this sense it is impossible to apply Berkeley's famous philosophical image of the tree in the quad. Does a sounded bell in the quad remain when nobody is there? One would have to answer, in the assumption the bell is only struck once: of course not, since sound fades by its own nature.

36 Cf. Rodolphe Gasché's still important work on Derrida in *The Tain of the Mirror: Derrida and the Philosophy of Reflection* (Cambridge, Mass.: Harvard University Press, 1986) and *Inventions of Difference: On Jacques Derrida* (Cambridge, Mass.: Harvard University Press, 1994).

37 G, p. 1a and 1b, the latter a quotation from Genet.

38 See both Critchley and Kimmerle for different elaborations of the way family figures throughout *Glas*, in *Hegel After Derrida*, pp. 198–212 and 233–237 respectively.

39 *Glas*, p. 4a.

40 Ibid., pp. 27–28a.

41 Ibid., p. 29a.

42 Ibid., p. 36a.

43 Ibid., p. 31a. Critchley takes this further in relation to the Holy Family and the mother, the latter of whom must drop out in the spiritual phallocentrism of the Father–Son relationship – *Hegel After Derrida*, pp. 205–207. Thomas Altizer takes this yet further, in his understanding of an apocalyptic Trinity, as we will see in our Conclusion.

44 *Glas*, p. 36a.

45 Gayatri Chakravorti Spivak, 'Glas-Piece: A Compte Rendu', *Diacritics* 7 (1977), 22–43. See also Michael Riffaterre, 'Syllepsis', *Critical Inquiry* 6 (1980), 625–638.

46 *Glas*, p. 77a. This would counter certain Gnostic readings of John's gospel, and likewise certain Neoplatonic readings.

47 We will see later how Jean-Luc Nancy focusses on God as principle to show how Christian monotheism deconstructs itself: a principle removes itself from any predication in materiality.

48 Ibid. It is here that we encounter the originating kernel of 'death of God' theology, which, again, we return to briefly in the end.
49 Ibid., p. 82a.
50 Ibid., pp. 111a and 100a respectively.
51 Ibid., p. 100b.
52 Ibid., p. 175b.
53 Cf. my 'Chapter 0: Introduction', in *Auden's O*.
54 Ibid., pp. 249a–250a.
55 *PS*, p. 424; *PG*, p. 512. Italics added.
56 *Glas*, p. 250a.
57 Ibid., p. 253a.
58 *PS*, p. 439.
59 *Glas*, pp. 261a–262a.
60 Ibid., p. 262a.
61 Ibid.
62 'It is remarkable that the deep mystery of tragedy and the deep mystery of Christianity are in such integral relationship with one another; in each atonement is the primal mystery, an atonement only possible through an absolute death or self-negation, one impossible for all gods and goddesses, but possible for both the tragic hero and incarnate Godhead' – Thomas J.J. Altizer, *The Apocalyptic Trinity* (New York: Palgrave Macmillan, 2012), pp. 136–137.
63 Ibid., p. 262.
64 The exception might be his Preface to Catherine Malabou's *The Future of Hegel: Plasticity, Temporality and Dialectic* (2005; orig. 1996). But see below, the end of Chapter 6, for comments on this Preface.

CHAPTER 6: THE LIVING OF HEGEL AND NEGATION: KRISTEVA, NANCY, AGAMBEN, ŽIŽEK, MALABOU

1 Though Kristeva had no recourse to *Glas*, she devotes a section to Derrida's *De la Grammatologie* (1967), and specifically to Derrida's *différance*, with its relation to negativity. Kristeva acknowledges that grammatology, with its attendant notions of difference, the trace, the grammè and writing (*écriture*), 'is, in our view, the most radical of all the various procedures that have tried, *after Hegel*, to push dialectical negativity further and elsewhere' (emphasis added) – *Revolution in Poetic Language*, trans. Margaret Waller (New York: Columbia University Press, 1984), p. 140. But she also is critical of *différance*'s capacity to produce and maintain the very kind of breaks and ruptures she has elaborated between the semiotic and the symbolic. See pp. 140–146 for the English abridgement of her argument. See also Juliana De Nooy, *Derrida, Kristeva, and the Dividing Line: An Articulation of Two Theories of Difference* (New York: Garland Publishing, 1998), for a further examination of the similarities and differences between the two thinkers.
2 *PS*, p. 35.
3 Ibid., p. 86.
4 Kristeva, *Revolution in Poetic Language*, p. 115. The relevant sections Kristeva here refers to and quotes from in the *Phenomenology* are §132–147 (pp. 79–89).
5 *Revolution in Poetic Language*, p. 16.
6 Ibid., p. 17.
7 See, for example, the Preface, p. 10, §18.

8   *PS*, p. 12.
9   *Revolution in Poetic Language*, p. 109.
10 In *Science of Logic*, *Negativität* is largely reserved for discussions of essence in Book Two, which one would expect if we understand negativity as more of a state than an action. Even so, that section, particularly in discussing illusory being and reflection, makes it clear that the activity of negation and the state of negativity cannot themselves be separated. The activity and the state are in a dialectical relation with one another. See especially pp. 397–401. For Kristeva, 'negation' still clings too closely to the Kantian sense of a real or logical opposition (*Revolution in Poetic Language*, pp. 117–118).
11 *Revolution in Poetic Language*, p. 109.
12 Within the context of the Absolute Idea, and in relation to the syllogistic movement of the negative with the *Begriff*, Hegel writes: 'As self-sublating contradiction this negativity is the *restoration* of the *first immediacy*, of simply universality; for the other of the other, the negative of the negative, is immediately the *positive*, the *identical*, the *universal*. If one insists on *counting*, the *second* immediate is, in the course of the method as a whole, the *third* term to the first immediate and the mediated. It is also, however, the third term of the first or formal negative and to absolute negativity or the second negative; now as the first negative is already the second term, the term reckoned as *third* can also be reckoned as *fourth*, and instead of a *triplicity* [*Triplizität*], the abstract form may be taken as a *quadruplicity* [*Quadruplizität*]' – p. 836; *WL-II*, p. 564.
13 For such disturbance, Kristeva is very much reliant upon Bataille here, especially his essay 'Hegel, Death and Sacrifice', as we explored above in relation to Derrida.
14 *Revolution in Poetic Language*, p. 113. For Kristeva, the 'thetic' represents the break in the *process* of signification when the motive forces that generate language solidify into a fixed position – a proposition, a judgement or a thesis. 'All enunciation,' she clarifies, 'whether of a word or of a sentence, is thetic' – p. 43.
15 Kristeva will nevertheless posit Bataille alongside Joyce as 'emblems of the most radical aspects of twentieth century literature' who move 'beyond madness and realism in a leap that maintains both "delirium" and "logic"' – ibid., p. 82.
16 *PS*, pp. 81–82, §136, e.g.
17 *Revolution in Poetic Language*, p. 117.
18 Ibid., p. 208. Kristeva elaborates this 'idea' of practice from Hegel's own idea of the 'practical Idea' in the latter stages of the *Science of Logic*, where the practical can only be fully realised once it has been absolutised in the Notion (*Begriff*). But the contradiction inherent in this realisation is precisely the contradiction that cannot be overcome by any absolutisation, because it *exceeds* the dialectal process – negativity as the fourth term. For Kristeva, therefore, the practical, as practice, must be part of a *material dialectic*, but one whose materiality is always heterogeneous to any unifying process. See ibid., pp. 195–207.
19 Ibid., p. 204.
20 Ibid., p. 113.
21 Ibid., p. 233.
22 As reprinted in *Desire in Language: A Semiotic Approach to Literature and Art*, ed. Leon S. Roudiez (New York: Columbia University Press, 1980), p. 93.
23 Ibid., p. 107.
24 Ibid., pp. 109, 110.
25 See *PS*, pp. 90–91, §148–149.
26 *Desire in Language*, p. 111.
27 *Powers of Horror: An Essay in Abjection*, trans. Leon S. Roudiez (New York: Columbia University Press, 1982).
28 *Revolution in Poetic Language*, p. 191.

29 Jean-Luc Nancy, *The Speculative Remark: (One of Hegel's Bons Mots)*, trans. Céline Surprenant (Stanford: Stanford University Press, 2001; orig. *La remarque spéculativ: (Un bon mot de Hegel)*, 1973).
30 Nancy acknowledges in the 'Preamble' that among those he takes for granted in his speculation (including Alexandre Koyré and Jean Hyppolite) stands Derrida and his two seminal essays on Hegel, 'From Restricted to General Economy: A Hegelianism Without Reserve' and 'The Pit and the Pyramid: Introduction to Hegel's Semiology' – ibid., p. 7.
31 Or more literally, a positionless unrest, with connotations of groundlessness and ceaselessness.
32 *SL*, p. 107.
33 *The Speculative Remark*, p. 38.
34 Ibid., p. 41.
35 Nancy, with Derridean precision, points out a slight difference in Hegel's text before the *Anmerkung*. Hegel (*SL*, p. 206) speaks of the unity of Being and Nothing as Becoming, but only as moments that are *vanishing* (*Verschwinden*). So when Becoming itself is overcome or suppressed, it too must vanish. Hegel uses the phrase 'vanishing of the vanishing' to describe the sublation of Becoming, rather than the term *aufheben* (or sublation of sublation). It seems that in the very sublation of itself, *Aufhebung* slides away and disappears in the text; it vanishes in the difference between *verschwinden* and *aufheben* (*The Speculative Remark*, p. 42), an instance of what Derrida might call *différance*, or Kristeva 'rejection' (and later 'abjection').
36 *The Speculative Remark*, p. 59.
37 Ibid., p. 60.
38 'Preface to the Second Edition', *SL*, pp. 31, 42.
39 *The Speculative Remark*, pp. 145–149.
40 Ibid., p. 97.
41 Ibid., p. 100, italics added.
42 Ibid., pp. 110ff.
43 *PS*, p. 491; *PG*, p. 588.
44 Jean-Luc Nancy, *Hegel: the Restlessness of the Negative*, trans. Jason Smith and Steven Miller (Minneapolis: University of Minnesota Press, 2002), p. 20.
45 *The Speculative Remark*, pp. 147–149.
46 Ibid., p. 148.
47 *Hegel: the Restlessness of the Negative*, p. 5. The English edition has added at the end key excerpts from Hegel's texts, the first of which is 'Thought as Effectivity', the three *Zusätze* from Section 19 of the *Lesser Logic*. There we find written that 'thinking is not an activity without content, for it produces thoughts and Thought itself' – *The Encyclopaedia Logic*, p. 48.
48 *Hegel: the Restlessness of the Negative*, p. 9.
49 Ibid., p. 21.
50 Ibid., p. 33.
51 Ibid., p. 36.
52 Ibid., p. 43.
53 Ibid., p. 50.
54 Ibid.
55 Ibid., p. 57.
56 Ibid., p. 75.
57 Ibid., p. 78. See Kristeva, *Powers of Horror*, pp. 1–2; and Emmanuel Levinas, *Entre Nous: Thinking-of-the-Other*, trans. Michael B. Smith and Barbara Harshav (New York: Columbia University Press, 1998): 'the in-itself of being-persisting-in-being goes

beyond itself in the gratuitousness of the outside-of-itself-for-the-other, in sacrifice' – p. xiii.
58 *Hegel: the Restlessness of the Negative*, p. 79; PS, p. 20.
59 Ibid., p. 54.
60 Trans. Philip Barnard and Cheryl Lester (New York: State University of New York Press, 1988).
61 Trans. Jeffrey S. Librett (Minneapolis: University of Minnesota Press, 1997).
62 Trans. Charlotte Mandell (New York: Fordham University Press, 2007).
63 Trans. Jeff Fort (New York: Fordham University Press, 2008).
64 Ed. Simon Sparks (Stanford: Stanford University Press, 2006).
65 *The Muses*, p. 36.
66 Ibid., p. 37. See also above, p. 17.
67 PS, pp. 58–66.
68 *Science of Logic*, p. 113. Agamben does not point out Heidegger's fundamental indebtedness to Hegel in relation to the term 'Dasein'. Perhaps this is because Agamben focusses on the *Phenomenology*; but even there, Hegel's sense is always of a situated existence, as opposed to abstract existence as such (*Existenz*). Cf. PG, p. 337, e.g. (PS, p. 309 – Miller is far from lucid in his translation here: 'existence' (*Existenz*) vs. 'actual existence' (*Dasein*). But then, Hegel himself was never perfectly clear, or consistent, on the difference. In *Science of Logic*, for example, within the second Remark on 'Becoming', he says, 'For being which is the outcome of *mediation* we shall reserve the term: *Existence*' (p. 93) – that is, being that is not in its immediate pure moment before it undergoes conversion by becoming. But Hegel will go on to argue that being can never remain in such a pure moment, since it is always giving way to its opposite. Defining such a pure moment is virtually impossible, since its 'purity' is the same as that of its opposite. Perhaps Heidegger, in his appropriation of 'Dasein', was signalling his attempt to isolate that very purity of Being, though not without its inherent negativity resident within – the *Dasein* with which Hegel's second Remark concludes (Miller's 'determinate being'), whereby Being can only be seen in relation to its opposite.)
69 Giorgio Agamben, *Language and Death: The Place of Negativity*, trans. Karen E. Pinkus and Michael Hardt (Minneapolis: University of Minnesota Press, 1991), p. 9. The poem itself is reprinted and retranslated on pp. 6–9.
70 PS, p. 63; *Language and Death*, p. 11.
71 *Language and Death*, p. 13.
72 Ibid., p. 9. ('Drum lebtest du auf ihrem Mund nicht. / Ihr Leben ehrte dich. In ihren Taten lebst du noch… / Dich ahn'ich oft als Seel ihrer Taten!' – p. 7.)
73 Cf. A-I, pp. 468–469, where the Mysteries are discussed under the section 'Affirmative Retention of Negated [*negativ gesetzten*] Features' (VA-II, Werke [Band] 14, pp. 64–66). By this stage, Hegel states that 'on the whole no great wisdom or profound knowledge seems to have been hidden in the Mysteries; on the contrary, they only preserved old traditions, the basis of what was later transformed by genuine art, and therefore they had for their contents not the true, the higher, and the better, but the lower and the more trivial.' Yet the very next sentence he acknowledges: 'This subject-matter, regarded as holy, was not clearly expressed in the Mysteries but was transmitted only in symbolical outlines. And in fact the character of the undisclosed and the unspoken belongs also to the old gods, to what was telluric, sidereal, and Titanic, for only the spirit is the revealed and the self-revealing. In this respect the symbolical mode of expression constitutes the other side of the secrecy in the Mysteries, because in the symbolical the meaning remains dark and contains something other than what the external, on which it is supposed to be represented, provides directly.'
74 *The Tempest*, IV.i.

75 *Language and Death*, p. 33. Agamben does not relate this intentionality here to Husserl, for unstated reasons. Perhaps he has Derrida's critique of Husserlian phenomenology in the back of his mind.
76 Ibid., p. 35; cf. p. 86: '*Death and Voice have the same negative structure and they are metaphysically inseparable.*'
77 Ibid., Epilogue, p. 108. '*Logic demonstrates that language is not my voice. The voice – it says – once was, but is no more nor can it ever be again. Language takes place in the non-place of the voice. This means that thought must think* nothing *in the voice*' – ibid.
78 Ibid., p. 90. This unusual but potent translation is not here acknowledged; we assume it is Agamben's. Fagles offers: 'Not to be born is best / when all is reckoned in, but once a man has seen the light / the next best thing by far, is to go back / back where he came from, quickly as he can.' – *Three Theban Plays*, trans. Robert Fagles (Harmondsworth: Penguin, 1982, 1984), p. 358 (ll. 1388–1391).
79 *Language and Death*, p. 90.
80 Agamben, *The Man Without Content*, p. 37.
81 Ibid., p. 49.
82 Ibid., p. 43.
83 Ibid.
84 *A-I*, p. 11.
85 Ibid., p. 9; *Man Without Content*, p. 52.
86 *Man Without Content*, p. 53.
87 Ibid.
88 Ibid., pp. 54, 55.
89 Ibid., p. 56.
90 Ibid., p. 57.
91 Ibid., p. 58.
92 We could compare this reading of art's traverse across modernity to the later versions of, say, Gianni Vattimo (*The End of Modernity*, orig. 1985) or Charles Taylor (*Sources of the Self*, 1989), both of whom have their Hegelian influence. Vattimo will open his text with 'nihilism as destiny', and speak frequently of the death of art as our destiny, expanding the question of aesthetic judgement to 'a general aestheticization of existence' (*The End of Modernity*, trans. Jon. R. Snyder (Cambridge: Polity Press, 1988), p. 52), even if Nietzsche and Heidegger remain the principal informants of this thesis. We might also compare Agamben to Jacques Rancière, and his attempt to 'rescue' aesthetics from its withered reputation, in *Aesthetics and Its Discontents*, trans. Steven Corcoran (Cambridge: Polity Press, 2009); and to Alain Badiou's collection of essays in *Handbook of Inaesthetics*, trans. Alberto Toscano (Stanford: Stanford University Press, 2005).
93 *Man Without Content*, p. 94, quoting from Bettina von Arnim, *Die Günderode*, vol. I (Leipzig: Insel, 1914), p. 331.
94 Ibid., p. 95.
95 Ibid., p. 99.
96 *Language and Death*, p. 95.
97 Žižek generally reads Marx (the Young or Left Hegelian) through Hegel; but he generally reads Freud through Lacan (the poststructuralist Freudian). One might argue that Lacan and Hegel, then, retain equal pre-eminence. However, in Žižek's recent tome, *Less Than Nothing: Hegel and the Shadow of Dialectical Materialism* (London: Verso, 2012), he explicitly reads Lacan through Hegel or as 'a repetition of Hegel' (p. 5), in a chapter entitled 'Lacan as a Reader of Hegel' within a section entitled 'The Thing Itself: Lacan'.
98 Slavoj Žižek, *First as Tragedy, Then as Farce* (London: Verso, 2009), p. 14.

99 Slavoj Žižek, *Looking Awry: An Introduction to Jacques Lacan through Popular Culture* (London: MIT Press, 1991), p. 159.
100 Slavoj Žižek, *The Fragile Absolute: Or, Why is the Christian Legacy Worth Fighting For?* (London: Verso, 1999), p. 46.
101 Slavoj Žižek, *On Belief* (London: Routledge, 2001), p. 110.
102 Slavoj Žižek, *The Ticklish Subject: The Absent Centre of Political Ontology* (London: Verso, 1999), p. 99. The original inversion of Popper, Žižek tells us, was put forward by Karel van het Reve, the Dutch linguist.
103 Ibid.
104 Slavoj Žižek, *The Puppet and the Dwarf: The Perverse Core of Christianity* (London: MIT Press, 2003), p. 91.
105 See the Conclusion below.
106 How would a publishing editor 'edit' Žižek? That is, in the interest of concision and a manageable/marketable page count, how would an editor go about choosing which of the myriad 'diversions' that constitute his meandering and circumlocutory prose to leave in and which to leave out? To select by ranking would be an impossible task: the entire edifice of prose, with all its manifold elaborations and deviations (theoretical equivalents, examples and counter-examples, anecdotes, film references, etc.), is an 'integrated diffusion', a paradox that has Hegel at its very root, as one side of the paradox negates the other, and as the ideas of argument are materialised in innumerable concrete manifestations, permutations and expressions. One can only let the edifice self-perpetuate as it will, even if, as with *Less than Nothing*, it distends to over a thousand pages!
107 *Ticklish Subject*, p. 77. Or as he says later in *Less Than Nothing* in relation to the Servant's 'overcoming' of the Bondsman: 'There is thus no reversal of negativity into positive greatness – the only 'greatness' here is this negativity itself' – p. 198.
108 *Ticklish Subject*, p. 80. He draws upon Vittorio Hösle for this idea of intersubjectivity. See p. 120, fn 8.
109 Ibid.
110 Ibid.
111 Ibid., p. 82; Beiser, *Hegel*, p. 201.
112 *Less Than Nothing*, p. 17.
113 Ibid., pp. 5–6.
114 Ibid., p. 333.
115 Cf. Heidegger's understanding of Hegel's negation, as explicated by Dahlstrom, in Introduction, note 14 above.
116 *Less Than Nothing*, p. 293, italics added.
117 Ibid., pp. 293–304.
118 Ibid., p. 304.
119 Ibid. It is not helpful that Žižek returns to the term 'negativity' now instead of 'negation'. It seems he wants here to use these two synonymously.
120 Ibid.
121 Robert Pippin, *The Persistence of Subjectivity: On the Kantian Aftermath* (Cambridge: Cambridge University Press, 2005), p. 281. An earlier version of this chapter ('What Was Abstract Art? (From the Point of View of Hegel)') was originally published in 2002 (*Critical Enquiry* 29, No. 1) and reproduced in Stephen Houlgate's edited volume *Hegel and the Arts* (Evanston: Northwestern University Press, 2007), pp. 244–270.
122 *Less Than Nothing*, p. 254. In postmodern art, Žižek says art's 'perversion is no longer subversive: the shocking excesses are part of the system itself, what the system feeds on in order to produce itself', so that the 'transgressive excess loses its shock value and is fully integrated into the established art market' – p. 256.

123 Ibid., pp. 611–612.
124 I have worked out this creative and circular paradox more extensively in *Auden's O: The Loss of One's Sovereignty in the Making of Nothing*.
125 Žižek will not use the example of Hegel himself but the example of the Christian God, with its Christology: 'our alienation from God coincides with the alienation of God from himself' – *Less Than Nothing*, p. 612. Again, we will see this further below in the theology of Thomas Altizer.
126 In the 'reflecting' of the gap, as the internal self-negation, the binary that might be construed in the shift from the transcendental reflection to the subjective reflection, as that between an absolute spirit and the subjectivisation of the object spirit, breaks down. Andrew Cutrofello has pointed out to me that Žižek sees this himself in ibid., pp. 98–99, but Cutrofello wonders here whether a third term of 'reflexion' might need to arise from the breakage. But the gap as negation is the overriding point: any third term of 'reflexion' that could arise must itself succumb to a breaking down: the gap is what characterises reflexion itself, as reflexivity.
127 Cf. *Hegel in označevalec* (Ljubljana: Univerzum, 1980); *Hegel in objekt*, with Mladen Dolar (Ljubljana: Društvo za teoretsko psihoanalizo, 1985).
128 See *A-I*, Section II, 'The Classical Form of Art', pp. 427ff, and *A-II*, Section II, 'Sculpture', pp. 701ff.
129 *A-II*, p. 719.
130 *A-I*, p. 594. A more elaborate account comes earlier: 'The beautiful feeling, the sentiment and spirit, of the happy harmony pervades all productions in which Greek freedom has become conscious of itself and portrayed its essence to itself. Therefore, the worldview of the Greeks is precisely the milieu in which beauty begins its true life and builds its serene kingdom; the milieu of free vitality which is not only there naturally and immediately but is generated by spiritual vision and transfigured by art; the milieu of a development of reflection and at the same time of that absence of reflection which neither isolates the individual nor can bring back to positive unity and reconciliation his negativity, grief, and misfortune; a milieu which yet, like life in general, is at the same time only a transitional point, even if at this point it attains the summit of beauty and in the form of its plastic individuality is so rich and spiritually concrete that all notes harmonize with it, and, moreover, what for its outlook is the past still occurs as an accessory and a background, even if no longer as something absolute and unconditioned' – p. 436.
131 Catherine Malabou, *The Future of Hegel*, p. 8. Italics added.
132 Ibid., p. 11.
133 Ibid., p. 69.
134 Ibid., pp. 11–12.
135 Ibid., p. 12.
136 Ibid., pp. 52–55.
137 Ibid., p. 119. We might compare, in light of both Aristotle's *Nous* and the death of God here, Thomas Altizer when talking about a *coincidentia oppositorum* 'only possible through a uniquely Hegelian *Aufhebung*': 'Here an Aristotelian *energia* is absolutely transformed, for Aristotle knows pure actuality or *energia* as the divine or metaphysical activity of pure thought (*nous*) eternally contemplating itself, a contemplation that is the contemplation of the Unmoved Mover, an Unmoved Mover which is the very opposite of absolute act. This is the classical metaphysical understanding that was revolutionized by Aquinas's discovery of Being or *esse* as that "act of being" lying at the root of the real as real, and thus it is identical with actual or real or pure existence. Now Aquinas can say what Aristotle could never say, that God is the pure Act-of-Being, and this "act of being" is identical with a pure actuality of being as being. If thereby a

138 Malabou, *The Future of Hegel*, p. 167.
139 *PS*, p. 36. As we saw in Chapter 1, Hegel speaks earlier in the Preface of this speculative movement as, following Aristotle, a self-movement of actuality (as opposed to potentiality or anticipation): the subject actualises itself through the movement that, in going out to the predicate, returns to itself in self-reflexivity, and in turn actualises the unity of both subject and predicate – *PS*, pp. 12–13, §23.
140 *Future of Hegel*, p. 179.
141 Ibid. The *trope* Malabou uses here to capture the turning back that, at the same time, projects in advance, is the French idiomatic phrase '*voir venir*', which the translator has rendered into English as 'to see (what is) coming'. In French the phrase 'means to wait, while, as is prudent, observing how events are developing. But it also suggests that other people's intentions and plans must be probed and guessed at. It is an expression that can thus refer at one and the same time to the state of "being sure of what is coming" ("être sûr de ce qui venir") and of "not knowing what is coming" ("ne pas savoir ce qui va venir")' – p. 13. Or later, '"To see (what is) coming" denotes at once the visibility and the invisibility of whatever comes... to see without seeing – await without awaiting – a future which is neither present to the gaze nor hidden from it' – p. 184. It is to await something new by dwelling in that something's past, suspended in the present between teleology and alienation, a suspension which determines the future as both beautiful and terrible: 'Beautiful because everything can still happen. Terrible because everything has already happened' – p. 192.
142 Ibid., p. xvi.
143 Ibid., p. 185.

CHAPTER 7: THE OUGHT OF NEGATION

1 *Philosophy of Right*, p. 23.
2 Ibid.
3 Ibid.
4 A fact even Kierkegaard could not fully emulate, despite, ironically, his best efforts.
5 *Philosophy of Right*, p. 26.
6 See for example Malabou's *What Should We Do with Our Brain?*, trans. Sebastian Rand (New York: Fordham University Press, 2008), or *The New Wounded: From Neurosis to Brain Damage*, trans. Steven Miller (New York: Fordham University Press, 2012), or the as yet untranslated *La chambre du milieu: De Hegel aux neurosciences* (Paris: Hermann, 2009). For a volume that combines all the domains of community, politics and religion, including Malabou's own contribution, see Žižek as co-editor: *Hegel and the Infinite*, eds Slavoj Žižek, Clayton Crockett and Creston Davis (New York: Columbia University Press, 2011).
7 The standard text to pursue all the intricacies and locations of Hegel's ethics is by one who edited Cambridge University Press's volume of *The Philosophy of Right* – Alan W. Wood, *Hegel's Ethical Thought* (Cambridge: Cambridge University Press, 1990).
8 *PS*, p. 373, §613, 614.
9 This is stated clearly at the beginning of Part III, 'Ethical Life', in *Philosophy of Right*, pp. 189–190, §148–149, which is then worked out in the remainder of the text.
10 Desmond, *Art and the Absolute*, pp. 99–100.

11 Recall Žižek's structural analysis of the *Greater Logic* as demanding a fourth term in the *Ticklish Subject*. See above, Chapter 6, pp. 139–20.
12 *SL*, p. 127.
13 Ibid.
14 *PS*, p. 126.
15 *SL*, p. 129.
16 Ibid.
17 Ibid., p. 132.
18 Hegel, in the *Anmerkung* (Remark) on this passage of the 'ought', unfolds the matter in relation to the question of possibility: '"You can, because you ought" – this expression, which is supposed to mean a great deal, is implied in the notion of ought. For the ought implies that one is superior to the limitation; in it the limit is sublated and the in-itself of the ought is thus an identical self-relation, and hence the abstraction of "can". But conversely, it is equally correct that: "you cannot, just because you ought." For in the ought, the limitation as limitation is equally implied; the said formalism of possibility has, in the limitation, a reality, a qualitative otherness opposed to it and the relation of each to the other is a contradiction, and thus a "cannot", or rather an impossibility' – ibid., p. 133.
19 Ibid., pp. 137–138.
20 Houlgate, *The Opening of Hegel's Logic*, p. 391.
21 Ibid., pp. 392–393.
22 *Philosophy of Right*, p. 41, §7. See also above, p. 187, n. 48.
23 See above, Chapter 6, and note 57.
24 Levinas, *Entre Nous*, p. 60.
25 See 'Violence and Metaphysics: An Essay on the Thought of Emmanuel Levinas', in *Writing and Difference*, pp. 79–153.
26 See *Liquid Modernity* (Oxford: Blackwell, 2000).
27 One thinks here principally of Ricoeur's ten studies that make up *Soi-même comme un autre*, or *Oneself as Another*, trans. Kathleen Blamey (Chicago: Chicago University Press, 1990).
28 *Revolution in Poetic Language*, p. 109.
29 Matthew 10:39 (KJV).
30 William Butler Yeats, 'Crazy Jane Talks to the Bishop'.
31 *SL*, p. 150. For the difference between 'bad' and 'spurious' in the translation of Hegel's original *schlechte Unendlichkeit*, see Wayne M. Martin, 'In Defense of Bad Infinity', in *Bulletin of the Hegel Society of Great Britain* 55/56 (2007), 168–187, esp. §1, 'The Logic of Bad Infinity'.
32 Jean-Luc Nancy, *Being Singular Plural*, trans. Robert D. Richardson and Anne E. O'Byrne (Stanford: Stanford University Press, 2000), pp. 182–183.
33 *Philosophy of Right*, pp. 41–42, §7 (Remark).
34 *The Man Without Content*, p. 55.
35 *Being Singular Plural*, p. 3.
36 'Why are There Several Art Forms?', in *The Muses*, p. 37.
37 *Oneself as Another*, p. 330.
38 *Hegel: The Restlessness of the Negative*, p. 60.

CONCLUSION: ART-RELIGION-PHILOSOPHY RE-FORMED

1 Søren Kierkegaard, *Fear and Trembling*, trans. Alastair Hannay (London: Penguin, 1985), p. 84.

2   Ibid., p. 109.
3   Jacques Derrida, *The Gift of Death*, trans. David Wills (Chicago: Chicago University Press, 1995), p. 59. This re-voices Kierkegaard himself: 'Silence hid in silence is suspicious, arouses mistrust, it is just as though one were to betray something, at least betrayed that one was keeping silence. But silence concealed by a decided talent for conversation – as true as ever I live, that is silence' – *The Journals of Kierkegaard*, ed. and trans. Alexander Dru (New York: Harper and Brothers, 1958, 1959), p. 245.
4   *Fear and Trembling*, p. 109.
5   Ibid.
6   It is true that, were Kierkegaard following the *Philosophy of Mind* only, he might justifiably conclude that the aesthetic is manifested as an immediacy, as in §556: 'As this consciousness of the Absolute first takes shape, its immediacy produces the factor of finitude in Art' – p. 293. Or in §557: (the God of) art as 'the so-called *unity* of nature of spirit – i.e. the immediate unity in sensuously intuitional form' – ibid. But in the earlier *Phenomenology*, the Religion of Art is a second stage following Natural Religion, or 'religion as *immediate*'. This second stage, manifested in the form of the self, is counterposed to the first, and only in the third stage, Revealed Religion, do the two opposing sides become reconciled (§683). And in the later *Aesthetics*, 'the sensuous aspect of the work of art, in comparison with the immediate existence of things in nature, is elevated to a pure appearance, and the work of art stands in the *middle* between immediate sensuousness and the ideal thought' – Introduction, p. 38.
7   *A-I*, p. 13.
8   Werner Hamacher, '(The End of Art with the Mask)', in *Hegel After Derrida*, ed. Stuart Barnett (London: Routledge, 1998), p. 105.
9   Ibid., p. 130.
10 Anton Chekhov, *About Love and Other Stories*, trans. Rosamund Bartlett (Oxford: Oxford University Press, 2004), p. 67.
11 Ibid., p. 76.
12 Ibid.
13 Bartlett has translated the Russian 'познания' (poznaniya) as 'consciousness', instead of the more common 'knowledge', or even 'cognition'. This might seem a significant deviation, until we keep in mind the Hegelian scheme that moves from consciousness to knowledge, at which point the translator's choice becomes (wilfully or unwilfully) prescient.
14 *The Journals of Kierkegaard*, p. 119.
15 *Godhead and the Nothing*, p. 127.
16 'The Transfiguration of Nothingness', in *The Movement of Nothingness*, p. 48.
17 Altizer, *The Apocalyptic Trinity*, p. 6.
18 Ibid., p. 17.
19 Ibid., p. 133.
20 Ibid., p. 134.
21 Altizer, 'The Transfiguration of Nothingness', p. 48.
22 As translated by David Farrell Krell, in *The Tragic Absolute: German Idealism and the Languishing of God* (Bloomington: Indiana University Press, 2005), pp. 22–26.
23 See Martin Gammon, 'Modernity and the Crisis of Aesthetic Representation in Hegel's Early Writings', in Maker, *Hegel and Aesthetics*, pp. 145–169, who argues that it was very unlikely Hegel was the author.
24 Krell, *The Tragic Absolute*, p. 2.
25 Ibid., p. 41.
26 See above, Chapter 6, pp. 130–131.

27 *The Tragic Absolute*, p. 43. See also p. 17, fn 2.
28 Agamben, *Language and Death*, p. 9.
29 *Fear and Trembling*, p. 147.

Bibliography of Works Cited

WORKS OF HEGEL (GERMAN)

Hegel, G.W.F. *Werke in zwanzig Bänden*. Eds E. Moldenhauer and K.M. Michel. Frankfurt: Suhrkamp, 1969–1971.
Werke [Band] 3, *Phänomenologie des Geistes* (1970).
Werke [Band] 5–6, *Wissenschaft Der Logik* (1969).
Werke [Band] 13–15, *Vorlesungen über die Ästhetik* (1970).
— *Dokumente zu Hegels Entwicklung*. Ed. J. Hoffmeister, 2nd ed. Stuttgart/Bad Cannstatt: Frommann-Holzboog, 1974.

WORKS OF HEGEL (ENGLISH)

Hegel, G.W.F. *The Philosophy of History*. Trans. J. Sibree. New York: Dover, 1956.
— *Science of Logic*. Trans. A.V. Miller. Amherst, NY: Humanity Books, 1969.
— *Philosophy of Mind*. Trans. William Wallace. Together with the *Zusätze* in Boumann's Text (1845), trans. A.V. Miller. Oxford: Clarendon Press, 1971.
— *Aesthetics: Lectures on Fine Art (Vols 1 and 2)*. Trans. T.M. Knox. Oxford: Oxford University Press, 1975.
— *The Difference Between Fichte's and Schelling's System of Philosophy*. Trans. H.S. Harris and Walter Cerf. Albany: SUNY Press, 1977.
— *Faith and Knowledge*. Trans. Walter Cerf and H.S. Harris. Albany: SUNY Press, 1977.
— *The Phenomenology of Spirit*. Trans. A.V. Miller. Oxford: Oxford University Press, 1977.
— *Elements of the Philosophy of Right*. Ed. Allen W. Wood. Trans. H.B. Nisbet. Cambridge: Cambridge University Press, 1991.
— *The Encyclopaedia Logic*. Trans. T.F. Geraets et al. Indianapolis: Hackett, 1991.
— *Hegel on Hamann*. Trans. Lisa Marie Anderson. Evanston: Northwestern University Press, 2008.

OTHER

Abrams, M.H. *Natural Supernaturalism: Tradition and Revolution in Romantic Literature*. New York: W.W. Norton, 1971.
Adorno, Theodor W. *Negative Dialectics*. Trans. E.B. Ashton. New York: Continuum, 2007.
Agamben, Giorgio. *Language and Death: The Place of Negativity*. Trans. Karen E. Pinkus and Michael Hardt. Minneapolis: University of Minnesota Press, 1991.
— *The Man Without Content*. Trans. Georgia Albert. Stanford: Stanford University Press, 1996.
— *State of Exception*. Trans. Kevin Attell. Chicago: Chicago University Press, 2005.
Altizer, Thomas J.J. *The Genesis of God: A Theological Genealogy*. Louisville: Westminster/John Knox Press, 1993.
— *Godhead and the Nothing*. Albany: SUNY Press, 2003.
— *The Apocalyptic Trinity*. New York: Palgrave Macmillan, 2012.
— 'The Transfiguration of Nothingness'. In *The Movement of Nothingness: Trust in the Emptiness of Time*. Eds Daniel Price and Ryan Johnson. Aurora: The Davies Group Publishers, 2013.
Aristotle. *De Anima*. Trans. Hugh Lawson-Tancred. London: Penguin, 1986.
— *Metaphysics*. Trans. Richard Hope. Ann Arbor: University of Michigan Press, 1952, 1960.
Badiou, Alain. *Handbook of Inaesthetics*. Trans. Alberto Toscano. Stanford: Stanford University Press, 2005.
Barnett, Stuart. 'Introduction'. In *Hegel after Derrida*. Ed. Stuart Barnett. London: Routledge, 1998.
Barnett, Stuart, ed. *Hegel after Derrida*. London: Routledge, 1998.
Bataille, Georges. *The Bataille Reader*. Eds Fred Botting and Scott Wilson. Oxford: Blackwell, 1997.
Bauman, Zygmunt. *Liquid Modernity*. Oxford: Blackwell, 2000.
Beiser, Frederick C. *Hegel*. New York: Routledge, 2005.
— 'Introduction: The Puzzling Hegel Renaissance'. In *The Cambridge Companion to Hegel and Nineteenth Century Philosophy*. Ed. Frederick C. Beiser. Cambridge: Cambridge University Press, 2008.
Beiser, Frederick C., ed. *The Cambridge Companion to Hegel and Nineteenth Century Philosophy*. Cambridge: Cambridge University Press, 2008.
Benjamin, Walter. *Illuminations*. Ed. H. Arendt. Trans. H. Zohn. New York: Schocken Books, 1968.
Bernstein, J.M. 'Freedom from Nature? Post-Hegelian Reflections on the End(s) of Art'. In *Hegel and the Arts*. Ed. Stephen Houlgate. Evanston: Northwestern University Press, 2007.
Blanchot, Maurice. *The Instant of My Death*. Trans. Elizabeth Rottenberg. Stanford: Stanford University Press, 2000.
Bungay, Stephen. *Beauty and Truth: A Study of Hegel's Aesthetics*. Oxford: Oxford University Press, 1984.

Butler, Judith. *Subjects of Desire: Hegelian Reflections in Twentieth-Century France*. New York: Columbia University Press, 1987.
Chekhov, Anton. *About Love and Other Stories*. Trans. Rosamund Bartlett. Oxford: Oxford University Press, 2004.
Comay, Rebecca. *Mourning Sickness: Hegel and the French Revolution*. Stanford: Stanford University Press, 2011.
Critchley, Simon. 'A Commentary Upon Derrida's Reading of Hegel in *Glas*'. In *Hegel After Derrida*. Ed. Stuart Barnett. London: Routledge, 1998.
Cutrofello, Andrew. 'Hamlet's Nihilism'. In *The Movement of Nothingness: Trust in the Emptiness of Time*. Eds Daniel Price and Ryan Johnson. Aurora: The Davies Group Publishers, 2013.
Dahlstrom, Daniel O. 'Thinking of Nothing: Heidegger's Criticism of Hegel's Conception of Negativity'. In *A Companion to Hegel*. Eds Stephen Houlgate and Michael Bauer. Oxford: Blackwell, 2011.
Danto, Arthur C. 'The Death of Art'. In *Death of Art*. Ed. Berel Lang. New York: Haven, 1984.
— *After the End of Art: Contemporary Art and the Pale of History*. Princeton: Princeton University Press, 1997.
de Boer, Karin. *Thinking in the Light of Time: Heidegger's Encounter with Hegel*. Albany: SUNY Press, 2000.
— *On Hegel: The Sway of the Negative*. Houndmills: Palgrave Macmillan, 2010.
de Nooy, Juliana. *Derrida, Kristeva, and the Dividing Line: An Articulation of Two Theories of Difference*. New York: Garland Publishing, 1998.
Derrida, Jacques. *Of Grammatology*. Trans. Gayatri Chakravorty Spivak. Baltimore: Johns Hopkins University Press, 1974, 1976.
— *Spurs: Nietzsche's Style; Éperons Les Styles de Nietzsche*. Trans. Barbara Harlow. Chicago: University of Chicago Press, 1978.
— *Writing and Difference*. Trans. Alan Bass. London: Routledge, 1978.
— *Dissemination*. Trans. Barbara Johnson. London: Athlone Press, 1981.
— *Positions*. Trans. Alan Bass. Chicago: University of Chicago Press, 1981; London: Athlone Press, 1987.
— *Margins of Philosophy*. Trans. Alan Bass. London: Harvester Wheatsheaf, 1982.
— *Glas*. Trans. John P. Leavey, Jr. and Richard Rand. Lincoln: University of Nebraska Press, 1986.
— *The Post Card: from Socrates to Freud and Beyond*. Trans. Alan Bass. Chicago: University of Chicago Press, 1987.
— *The Truth in Painting*. Trans. Geoff Bennington and Ian McLeod. Chicago: University of Chicago Press, 1987.
— 'Otobiographies: The Teaching of Nietzsche and the Politics of the Proper Name'. In *The Ear of the Other: Otobiography, Transference, Translation*. Ed. Christie McDonald. Trans. Peggy Kamuf and Avital Ronell. Lincoln: University of Nebraska Press, 1988.

— *The Gift of Death*. Trans. David Wills. Chicago: Chicago University Press, 1995.
— *Demeure: Fiction and Testimony*. Trans. Elizabeth Rottenberg. Stanford: Stanford University Press, 2000.
Desmond, William. *Art and the Absolute: A Study of Hegel's Aesthetics*. Albany: SUNY Press, 1986.
— 'Art and the Absolute Revisited: The Neglect of Hegel's Aesthetics'. In *Hegel and Aesthetics*. Ed. William Maker. Albany: SUNY Press, 2000.
Donougho, Martin. 'Art and History; Hegel on the End, the Beginning, and the Future of Art'. In *Hegel and the Arts*. Ed. Stephen Houlgate. Evanston: Northwestern University Press, 2007.
Fichte, J.G. *The Science of Knowledge*. Ed. and trans. Peter Heath and John Lachs. Cambridge: Cambridge University Press, 1982.
Flynn, Thomas. *Sartre, Foucault, and Historical Reason, Volume One: Toward an Existentialist Theory of History*. Chicago: University of Chicago Press, 1997.
Forster, Michael N. *Hegel's Idea of a Phenomenology of Spirit*. Chicago: University of Chicago Press, 1998.
Foucault, Michel. *The Archaeology of Knowledge*. Trans. A.M. Sheridan Smith. New York: Pantheon, 1972.
Fukuyama, Francis. *The End of History and the Last Man*. London: Penguin, 1992.
Gammon, Martin. 'Modernity and the Crisis of Aesthetic Representation in Hegel's Early Writings'. In *Hegel and Aesthetics*. Ed. William Maker. Albany: SUNY Press, 2000.
Gasché, Rodolphe. *The Tain of the Mirror: Derrida and the Philosophy of Reflection*. Cambridge, Mass.: Harvard University Press, 1986.
— *Inventions of Difference: On Jacques Derrida*. Cambridge, Mass.: Harvard University Press, 1994.
Goethe, Johann Wolfgang von. *Goethe's Faust*. Trans. Walter Kaufmann. New York: Anchor Books, 1961.
Habermas, Jürgen. *The Philosophical Discourse of Modernity*. Trans. Frederick Lawrence. Oxford: Polity Press, 1987.
Hahn, Sonsuk Susan. *Contradiction in Motion: Hegel's Organic Concept of Life and Value*. Ithaca: Cornell University Press, 2007.
Halper, Edward. 'The Logic of Art: Beauty and Nature'. In *Hegel and Aesthetics*. Ed. William Maker. Albany: SUNY Press, 2000.
Hamacher, Werner. '(The End of Art with the Mask)'. In *Hegel After Derrida*. Ed. Stuart Barnett. London: Routledge, 1998.
Hass, Andrew W. *Poetics of Critique: The Interdisciplinarity of Textuality*. Aldershot: Ashgate, 2003.
— 'Artist Bound: The Enslavement of Art to the Hegelian Other'. *Literature and Theology* 25/4 (2011), 379–392.
— *Auden's O: The Loss of One's Sovereignty in the Making of Nothing*. Albany: SUNY Press, 2013.

— 'Becoming'. In *Resurrecting the Death of God: The Past, Present, and Future of Radical Theology*. Eds Daniel Peterson and Michael Zbaraschuk. Albany: SUNY Press, 2014.
Heidegger, Martin. *Hegel's Phenomenology of Spirit*. Trans. Parvis Emad and Kenneth Maly. Bloomington: Indiana University Press, 1988.
— 'Hegel: Die Negativität. Eine Auseinandersetzung mit Hegel aus dem Ansatz in der Negativität (1938/39, 1941)'. In *Gesamtausgabe, Band 68*. Ed. lngrid Schüssler. Frankfurt am Main: Klostermann, 1993.
— 'What is Metaphysics?'. Trans. David Farrell Krell. In *Pathmarks*. Ed. William McNeill. Cambridge: Cambridge University Press, 1998.
Henrich, Dieter. 'Art and Philosophy of Art Today: Reflections with Reference to Hegel'. In *New Perspectives in German Literary Criticism: A Collection of Essays*. Eds R.E. Amacher and V. Lange. Princeton: Princeton University Press, 1979.
Hodgson, Peter C. *G.W.F. Hegel: Theologian of the Spirit*. London: T&T Clark, 1997.
— *Hegel and Christian Theology: A Reading of the Lectures on the Philosophy of Religion*. Oxford: Oxford University Press, 2005.
Houlgate, Stephen. 'Hegel and the "End" of Art'. *Owl of Minerva*, 29/1 (1997), 1–21.
— *The Opening of Hegel's Logic: From Being to Infinity*. West Lafayette, IN: Purdue University Press, 2006.
— 'Hegel's Logic'. In *The Cambridge Companion to Hegel and Nineteenth Century Philosophy*. Ed. Frederick C. Beiser. Cambridge: Cambridge University Press, 2008.
— 'Essence, Reflexion and Immediacy in Hegel's *Science of Logic*'. In *A Companion to Hegel*. Eds Stephen Houlgate and Michael Bauer. Oxford: Blackwell, 2011.
Houlgate, Stephen, ed. *Hegel and the Arts*. Evanston: Northwestern University Press, 2007.
Hyppolite, Jean. *Genesis and Structure of Hegel's Phenomenology of Spirit*. Trans. Samuel Cherniak and John Heckman. Evanston: Northwestern University Press, 1974.
— *Logic and Existence*. Trans. Leonard Lawlor and Amit Sen. Albany: SUNY Press, 1997.
Inwood, Michael. *Hegel*. London: Routledge, 1983, 2002.
James, David. *Art, Myth and Society in Hegel's Aesthetics*. London: Continuum, 2009.
Jameson, Fredric. *The Hegel Variations: On the Phenomenology of the Spirit*. London: Verso, 2010.
Kaminsky, Jack. *Hegel on Art: An Interpretation of Hegel's Aesthetics*. Albany: SUNY Press, 1962.
Kant, Immanuel. *Critique of Pure Reason*. Trans. Norman Kemp Smith. Houndmills: Macmillan, 1933.
— *Critique of Judgment*. Trans. James Creed Meredith. Oxford: Clarendon Press, 1952.
Kierkegaard, Søren. *The Journals of Kierkegaard*. Ed. and trans. Alexander Dru. New York: Harper and Brothers, 1958, 1959.
— *Fear and Trembling*. Trans. Alastair Hannay. London: Penguin, 1985.
Kimmerle, Heinz. 'On Derrida's Hegel Interpretation'. In *Hegel after Derrida*. Ed. Stuart Barnett. London: Routledge, 1998.

Kirk, G.S., J.E. Raven and M. Schofield, eds. *The Presocratic Philosophers, 2nd Ed*. Cambridge: Cambridge University Press, 1957, 1983.

Kolb, David. *The Critique of Pure Modernity: Hegel, Heidegger and After*. Chicago: Chicago University Press, 1986.

Kojève, Alexandre. *Introduction to the Reading of Hegel*. Ed. Allan Bloom. Trans. James H. Nichols, Jr. New York: Cornell University Press, 1969.

Krell, David Farrell. *The Tragic Absolute: German Idealism and the Languishing of God*. Bloomington: Indiana University Press, 2005.

Kristeva, Julia. *Desire in Language: A Semiotic Approach to Literature and Art*. Ed. Leon S. Roudiez. New York: Columbia University Press, 1980.

— *Powers of Horror: An Essay in Abjection*. Trans. Leon S. Roudiez. New York: Columbia University Press, 1982.

— *Revolution in Poetic Language*. Trans. Margaret Waller. New York: Columbia University Press, 1984.

Lacan, Jacques. *Écrits: A Selection*. Trans. Alan Sheridan. New York: Norton, 1977.

Lang, Berel, ed. *Death of Art*. New York: Haven, 1984.

Lauer, Quentin. *A Reading of Hegel's Phenomenology of Spirit*. New York: Fordham University Press, 1976, 1993.

Leavey, John P. Jr. *Glassary*. Lincoln: University of Nebraska Press, 1986.

Levinas, Emmanuel. *Entre Nous: Thinking-of-the-Other*. Trans. Michael B. Smith and Barbara Harshav. New York: Columbia University Press, 1998.

Mahon, Peter. *Imagining Joyce and Derrida: Between Finnegans Wake and Glas*. Toronto: University of Toronto Press, 2007.

Maker, William, ed. *Hegel and Aesthetics*. Albany: SUNY Press, 2000.

Malabou, Catherine. *The Future of Hegel: Plasticity, Temporality and Dialectic*. Trans. Lisabeth During. Abingdon: Routledge, 2005.

— *What Should We Do with Our Brain?* Trans. Sebastian Rand. New York: Fordham University Press, 2008.

— *La chambre du milieu: De Hegel aux neurosciences*. Paris: Hermann, 2009.

— *The New Wounded: From Neurosis to Brain Damage*. Trans. Steven Miller. New York: Fordham University Press, 2012.

Martin, Wayne M. 'In Defense of Bad Infinity'. *Bulletin of the Hegel Society of Great Britain* 55/56 (2007), 168–187.

Marx, Karl. *The 18th Brumaire of Louis Bonaparte*. New York: International Publishers, 1963.

Melamed, Yitzhak Y. '"Omnis determinatio est negatio": determination, negation, and self-negation in Spinoza, Kant, and Hegel'. In *Spinoza and German Idealism*. Eds Eckart Förster and Yitzhak Y. Melamed. Cambridge: Cambridge University Press, 2012.

Morani, Roberto. 'Heidegger, Hegel und die Frage des Nichts'. In *Nichts – Negation – Nihilismus: Die europäische Moderne als Erkenntnis und Erfahrung des Nichts*. Eds Allessandro Bertinetto and Christoph Binkelmann. Frankfurt am Main: Peter Lang, 2010.

Nancy, Jean-Luc. *The Literary Absolute: The Theory of Literature in German Romanticism*. Trans. Philip Barnard and Cheryl Lester. New York: State University of New York Press, 1988.
— *The Muses*. Trans. Peggy Kamuf. Stanford: Stanford University Press, 1996.
— *The Sense of the World*. Trans. Jeffrey S. Librett. Minneapolis: University of Minnesota Press, 1997.
— *Being Singular Plural*. Trans. Robert D. Richardson and Anne E. O'Byrne. Stanford: Stanford University Press, 2000.
— *The Speculative Remark: (One of Hegel's Bons Mots)*. Trans. Céline Surprenant. Stanford: Stanford University Press, 2001.
— *Hegel: the Restlessness of the Negative*. Trans. Jason Smith and Steven Miller. Minneapolis: University of Minnesota Press, 2002.
— *Multiple Arts: The Muses II*. Ed. Simon Sparks. Stanford: Stanford University Press, 2006.
— *Listening*. Trans. Charlotte Mandell. New York: Fordham University Press, 2007.
— *The Ground of the Image*. Trans. Jeff Fort. New York: Fordham University Press, 2008.
Nietzsche, Friedrich. *The Gay Science*. Ed. Bernard Williams. Trans. Josefine Nauckhoff Cambridge: Cambridge University Press, 2001.
Nuzzo, Angelica. 'How Does Nothing(ness) Move? Hegel's Challenge to Embodied Thinking'. In *The Movement of Nothingness: Trust in the Emptiness of Time*. Eds Daniel Price and Ryan Johnson. Aurora: The Davies Group Publishers, 2013.
Pippin, Robert B. 'What Was Abstract Art? (From the Point of View of Hegel)'. In *Critical Inquiry* 29/1 (2002), 575–598.
— *The Persistence of Subjectivity: On the Kantian Aftermath*. Cambridge: Cambridge University Press, 2005.
— 'What Was Abstract Art? (From the Point of View of Hegel)'. In Stephen Houlgate, ed. *Hegel and the Arts*. Evanston: Northwestern University Press, 2007.
— 'The Absence of Aesthetics in Hegel's Aesthetics'. In *The Cambridge Companion to Hegel and Nineteenth Century Philosophy*. Ed. Frederick C. Beiser. Cambridge: Cambridge University Press, 2008.
Price, Daniel and Ryan Johnson, eds. *The Movement of Nothingness: Trust in the Emptiness of Time*. Aurora: The Davies Group Publishers, 2013.
Rancière, Jacques. *Aesthetics and Its Discontents*. Trans. Steven Corcoran. Cambridge: Polity Press, 2009.
Ricoeur, Paul. *Hermeneutics and the Human Sciences*. Ed. and trans. John B. Thompson. Cambridge: Cambridge University Press, 1981.
— *Oneself as Another*. Trans. Kathleen Blamey. Chicago: Chicago University Press, 1990.
Riffaterre, Michael. 'Syllepsis'. *Critical Inquiry* 6 (1980), 625–638.
Roughley, Alan. *Reading Derrida Reading Joyce*. Gainsville: University Press of Florida, 1999.
Rutter, Benjamin. *Hegel on the Modern Arts*. Cambridge: Cambridge University Press, 2010.

Sartre, Jean-Paul. *What is Literature? And Other Essays*. Trans. Bernard Frechtman, et al. Cambridge, Mass.: Harvard University Press, 1988.
— *Being and Nothingness*. Trans. Hazel E. Barnes. London: Routledge, 2003.
— *Critique of Dialectical Reason, Vol. I*. Trans. Alan Sheridan-Smith. London: Verso, 1991, 2004.
— *Critique of Dialectical Reason, Vol. II*. Trans. Quintin Hoare. London: Verso, 1991, 2006.
Schelling, F.W.J. *System of Transcendental Idealism (1800)*. Trans. Peter Heath. Charlottesville: University Press of Virginia, 1978.
— *Philosophical Investigations into the Essence of Human Freedom*. Trans. Jeff Love and Johannes Schmidt. Albany: SUNY Press, 2006.
Schiller, Friedrich. 'Über die ästhetische Erziehung des Menschen, in einer Reihe von Briefen'. In *Schillers sämtliche Werke, vierter Band*. Stuttgart: J.G. Cotta'sche Buchhandlung, 1879.
— *On the Aesthetic Education of Man*. Trans. Reginald Snell. Mineola, NY: Dover, 2004.
Shakespeare, William. *The Complete Works*. Gen. ed. Alfred Harbage. New York: Viking, 1969.
Sophocles. *Oedipus at Colonus*. In *Three Theban Plays*. Trans. Robert Fagles. Harmondsworth: Penguin, 1982, 1984.
Speight, Allen. *Hegel, Literature and the Problem of Agency*. Cambridge: Cambridge University Press, 2001.
— 'Hegel's Aesthetics: The Practice and "Pastness" of Art'. In *The Cambridge Companion to Hegel and Nineteenth Century Philosophy*. Ed. Frederick C. Beiser. Cambridge: Cambridge University Press, 2008.
Spivak, Gayatri Chakravorti. 'Glas-Piece: A Compte Rendu'. *Diacritics* 7 (1977), 22–43.
Steinkraus, Warren E. and Kenneth L. Schmitz, eds. *Art and Logic in Hegel's Philosophy*. Sussex: Harvester Press, 1980.
Stewart, Jon. 'Hegel's *Phenomenology* as a Systematic Fragment'. In *The Cambridge Companion to Hegel and Nineteenth Century Philosophy*. Ed. Frederick C. Beiser. Cambridge: Cambridge University Press, 2008.
Sussman, Henry. 'Hegel, *Glas*, and the Broader Modernity'. In *Hegel after Derrida*. Ed. Stuart Barnett. London: Routledge, 1998.
Taylor, Charles. *Hegel*. Cambridge: Cambridge University Press, 1975.
— *Sources of the Self*. Cambridge, Mass.: Harvard University Press, 1989.
Vater, Michael. 'Introduction'. In F.W.J. Schelling, *System of Transcendental Idealism (1800)*. Trans. Peter Heath. Charlottesville: University Press of Virginia, 1978.
Vattimo, Gianni. *The End of Modernity*. Trans. Jon. R. Snyder. Cambridge: Polity Press, 1988.
Wicks, Robert. *Hegel's Theory of Aesthetic Judgment*. New York: Peter Lang, 1994.
Wood, Alan W. *Hegel's Ethical Thought*. Cambridge: Cambridge University Press, 1990.
Wyss, Beat. *Hegel's Art History and the Critique of Modernity*. Trans. Caroline Dobson Saltzwedel. Cambridge: Cambridge University Press, 1999.
Žižek, Slavoj. *Hegel in označevalec*. Ljubljana: Univerzum, 1980.
— *Looking Awry: An Introduction to Jacques Lacan through Popular Culture*. London: MIT Press, 1991.

— *The Fragile Absolute: Or, Why is the Christian Legacy Worth Fighting For?* London: Verso, 1999.
— *The Ticklish Subject: The Absent Centre of Political Ontology.* London: Verso, 1999.
— *On Belief.* London: Routledge, 2001.
— *The Puppet and the Dwarf: The Perverse Core of Christianity.* London: MIT Press, 2003.
— *First as Tragedy, Then as Farce.* London: Verso, 2009.
— *Less Than Nothing: Hegel and the Shadow of Dialectical Materialism.* London: Verso, 2012.
Žižek, Slavoj, and Mladen Dolar. *Hegel in object.* Ljubljana: Društvo za teoretsko psihoanalizo, 1985.
Žižek, Slavoj, Clayton Crockett and Creston Davis, eds. *Hegel and the Infinite: Religion, Politics and Dialectic.* New York: Columbia University Press, 2011.

Index

Italicised pages denote subjects with sustained emphasis

Abrams, M.H., 187 n. 2
Adorno, Theodor W., 77
Agamben, Giorgio, 17–18, 36–37, *127–135*, 136, 143, 146, 157, 164, 165, 166, 180, 181
Althusser, Louis, 143
Altizer, Thomas J.J., 4, 18, 138, *176–178*, 179, 201 n. 43, 202 n. 62, 208 n. 124, 208 n. 137
apophaticism, 8
Aquinas, Thomas, 107, 208 n. 137
Aristotle, 12, 33, 47, 107, 115, 149, 185 n. 25, 190 n. 11, 208 n. 137, 208 n. 139. Aristotelianism, 26, 52
Arnim, Bettina von, 206 n. 93
Artaud, Antonin, 119

Badiou, Alain, 4, 206 n. 92
Barnett, Stuart, 200 n. 7
Barthes, Roland, 118
Bataille, Georges, 92, *93–97*, 100, 103, 109, 113, 115, 116, 119, 135, 200 n. 18, 203 n. 13, 203 n. 15
Bauman, Zygmunt, 164
Beckett, Samuel, 178
becoming, 7, 10, 27, 31, 33, 47, *50–53*, 80, 84, 96, 120–121, 123, 125, 126, 193 n. 51, 193 n. 52, 198 n. 7, 200 n. 19, 204 n. 35, 205 n. 68
Beiser, Frederick, 4, 6, 24, 140
Benjamin, Walter, 134, 188 n. 25
Berkeley, George, 201 n. 35
Bernstein, J.M., 197 n. 49
Blanchot, Maurice, 51, 97, 113, 193 n. 48
Borges, Jorge Luis, 119
Brandom, Robert, 184 n. 11
Bungay, Stephen, 16, 196 n. 14, 196 n. 24
Butler, Judith, 9, 85, 184 n. 15, 198 n. 4

Céline, Louis-Ferdinand, 119
Chaplin, Charlie, 139
Chekhov, Anton, 173–176, 178, 181–182
Chesterton, G.K., 143
cogenitive, cogenitivity, *44–47*, 51, 52, 53, 68, 69, 72, 73, 89, 90, 93, 104, 105, 113, 120, 121, 124, 136, 140, 146, 148, 150–151, 156, 160, 166, 171, 176, 191 n. 21, 191 n. 22, 192 n.43
Comay, Rebecca, 184 n. 17
Critchley, Simon, 201 n. 31, 201 n. 33, 201 n. 38, 201 n. 43

224 INDEX

Crockett, Clayton, 209 n. 6
Cutrofello, Andrew, 183–184 n. 5, 208 n. 126

Dahlstrom, Daniel O., 184 n. 14, 207 n. 115
Danto, Arthur C., 16, 188–189 n. 6
Davis, Creston, 209 n. 6
death of God, 4, 149, *176–178*, 202 n. 48
de Boer, Karin, 187 n. 1, 190 n. 11, 193 n. 57
deconstruction, 78, 95, 96, 100
Deleuze, Gilles, 85, 144
de Nooy, Juliana, 202 n. 1
Derrida, Jacques, 17, 83, 85, *91–112*, 113, 114, 120, 121, 128, 145, 151, 156, 164, 170, 176, 180, 181, 188 n. 21, 193 n.48, 202 n. 1, 203 n. 13, 204 n. 30, 204 n. 35, 206 n. 75
Descartes, Rene, 13, 46
    Cartesianism, 5, 27, 42, 48
    cogito, 5, 42, 139.
Desmond, William, 16, 59, 158, 190 n. 14, 191 n. 25, 192 n. 31, 196 n. 14, 196 n. 17, 196 n. 24
Donougho, Martin, 197 n. 49
Dostoyevsky, Fyodor, 119

end of art, death of art, 16–19, 36, 57, *70–73*, 78, *132–135*, 206 n. 92
existential, existentialism, 15, 78, 79, 82, 85, 128, 186 n. 34

Feuerbach, Ludwig, 92, 113
Fichte, J.G., 5, 11, 12, 13, 185 n. 28, 187 n. 48, 191 n. 18, 196 n. 32
Forster, Michael N., 188 n. 13

Foucault, Michel, 82, 85, 90
fourth term, 116, 119, 140, 141–142, 165, 203 n. 12, 203 n. 18, 210 n. 11
Freud, Sigmund, 14, 27, 114, 119, 125, 136, 143, 206 n. 97
Fukuyama, Francis, 4, 184 n. 6

Gadamer, Hans-Georg, 183 n. 4
Gammon, Martin, 57, 211 n. 23
Gasché, Rodolphe, 194 n. 66, 201 n. 36
Genet, Jean, 17, 92, 102, 107, 108–109, 112
Goethe, Johann Wolfgang von, 12, 185 n. 28
Gnosticism, 201 n. 46
Greenberg, Clement, 188–189 n. 6

Habermas, Jürgen, 13, 19
Hahn, Sonsuk Susan, 194 n. 66
Halper, Edward, 196 n. 24
Hamacher Werner, 16, 172
Hamann, Johann Georg, 187 n. 7
Hartmann, Klaus, 184 n. 11
Hass, Andrew W., 184 n. 18, 194 n. 67, 197 n. 49, 198 n. 2, 198 n. 19, 200 n. 20, 202 n. 53, 208 n. 124
Hegelianism(s), 1, 2–3, 5, 7, 9, 18, 19, 23, 30, 36, 82, 83, 84, 90, 92, 93–94, 124, 146, 158, 165, 169, 170, 178, 182
Heidegger, Martin, 37, 42, 46, 47, 53, 79, 113, 128, 160, 184 n. 14, 190 n. 11, 193 n. 54, 205 n. 68, 206 n. 92, 207 n. 115
Heine, Heinrich, 143
Henrich, Dieter, 16
Heraclitus, 190 n. 11
hermeneutics, 3, 149, 150–151, 166, 183 n. 4, 196 n. 32, 199–200 n. 6

Hesiod, 47
Hodgson, Peter, 4
Hölderlin, Friedrich, 14, 128, 134, 179–180, 181, 182
Homer, 47
Hösle, Vittorio, 207 n. 108
Hotho, H.G., 194 n. 4
Houlgate, Stephen, 16, 44, 161, 162, 190 n. 12, 193 n. 58, 194 n. 66, 198 n. 10
Husserl, Edmund, 48, 79, 206 n. 75
Hyppolite, Jean, 83, *85–90*, 92, 113, 184 n. 15, 199 n. 41, 204 n. 30

Inwood, Michael, 192 n. 26

Jacobi, Friedrich Heinrich, 5, 192 n. 39, 193 n. 44
James, David, 16
Jameson, Fredric, 188 n. 13
Jesus Christ, 190 n. 11
Joyce, James, 101, 111, 119, 203 n. 15
Jung, Carl, 143

Kafka, Franz, 134
Kaminsky, Jack, 16
Kant, Immanuel, 5, 12, 13, 19, 23, 26, 37, 41, 43, 44, 45, 54, 61, 62, 63–64, 83, 96, 131, 138, 144, 158, 191 n. 18, 192 n. 39, 194 n. 6, 195 n. 8, 196 n. 27
    Kantianism, 5, 12, 13, 40, 41, 42, 43, 56, 63, 127, 157, 203 n. 10
kenosis, 34, 35, 37, 110–111, 112, 133, 135, 149, 177, 178
Kierkegaard, Søren, 69, 78, 92, 102, 113, 127, *169–171*, 172, 174–175, 176, 178, 181, 182, 201 n. 33, 209 n. 4, 211 n. 3
Kimmerle, Heinz, 200 n. 10, 201 n. 38

Kojève, Alexandre, 3, 4, 79, 85, 89, 92, 94, 96, 113, 184 n. 6, 200 n. 19
Kolb, David, 190 n. 11
Koyré, Alexandre, 204 n. 30
Krell, David Farrell, *179–181*
Kristeva, Julia, 91, *113–119*, 120, 121, 122, 125, 126, 128, 136, 140, 146, 157, 164, 165, 166, 202 n. 1, 204 n. 35

Lacan, Jacques, 14, 79, 136, 139, 141–142, 144, 186 n. 34, 206 n. 97
Lacoue-LaBarthe, Philippe, 126–127
Lang, Berel, 16
Lauer, Quentin, 190 n. 9
Lautréamont, Comte de, 117
Leavey, John P. Jr., 201 n. 31
Leibniz, Gottfried Wilhelm, 186 n. 33
Leiris, Michel, 98–100, 102
Levinas, Emmanuel, 126, 162–164

Mahon, Peter, 200 n. 30
Maker, William, 16
Malabou, Catherine, *145–152*, 155, 157, 164, 165, 166, 184 n. 13, 189 n. 33, 202 n. 64, 209 n. 6
Mallarmé, Stéphane, 117
Martin, Wayne M., 210 n. 31
Marx, Karl, 4, 9, 92, 94, 113, 143, 156, 183 n. 2, 206 n. 97
    Ideologiekritik, 78
    Marxism, 4, 136, 156, 178, 198 n. 21
master/slave, 14, 70, *94–95*, 97, 197 n.49, 200 n. 20
Melamed, Yitzhak Y., 193 n. 55
metaphor, 8, *25–26*, 28, 40, 87, 90, 99, 141, 188 n. 9
Milton, John, 177
Morani, Roberto, 913 n. 54

Nancy, Jean-Luc, 17, 18, 37, 51, 52, *119–127*, 128, 130, 132, 134, 136, 146, 157, 165, 166–167, 201 n. 47
Naturphilosophie, philosophy of nature, 4
Neoplatonism, 23, 201 n. 46
Nietzsche, Friedrich, 14, 37, 41, 77–78, 122, 129, 134, 135, 165, 181, 186 n. 33, 206 n. 92
nihilism, 7, 9, 17, 18, 77–78, 133, 135, 183 n. 5, 206 n. 92
Nuzzo, Angelica, 10

organicism, 12, 25–26, 28, 37, 40, 42, 43, 47, 55, 58, 69, 90, 131, 161–162, 190 n. 14

Parmenides, vii, 8, 49, 81, 170, 190 n. 11
Pippin, Robert, 16, 58, 144, 156, 184 n. 11, 194 n. 4, 195–196 n. 14
Plato, Platonism, 23, 46, 83, 87–88, 190 n. 11
Popper, Karl, 137
post-metaphysics, 5, 6–7, 17, 18
postmodernity, postmodernism, 6, 14, 37, 41, 156–157, 162, 176
Presocratics, 47
Proust, Marcel, 119
Pythagoras, Pythagoreanism, 47

Rancière, Jacques, 206 n. 92
reflection, philosophy of reflection, 5, 6, 53, 54, 57, 62, 69, 70, 86, 104, 131, 194 n. 66, 196 n. 27, 208 n. 126, 208 n. 130, 208 n. 139
Reve, Karel van het, 207 n. 102
Ricoeur, Paul, 3, 164, 166, 183 n.4, 210 n. 27
Riffaterre, Michael, 201, n. 45

Rilke, Rainer Maria, 52
Romanticism, 14–15, 61
Rothko, Mark, 41, 156
Roughley, Alan, 200 n. 30
Rutter, Benjamin, 16

Sartre, Jean-Paul, *78–85*, 86, 89, 92, 105, 113, 142, 198 n. 4, 198–199 n. 29
Schelling, F.W.J., 12, 13, 14–15, 26, 179–180
Schiller, Friedrich, 14, 26, *27–29*, 30, 31–32, 35, 36, *57–58*, 61, *62–65*, 66, 68, 146, 147–148, 188 n. 9
Schmitz, Kenneth L., 16
Shaftesbury, 3rd Earl of, 131
Shakespeare, William, 29, 129–130, pp. 183–184 n. 5
shape, *Gestalt*, shaping, 27–30, 31, 32, 34, 35–36, 56, 59, 62, 65, 66, 68, 69, 146, 147
Sophocles, 26, 130
sovereignty, 95–97, 103, 135, 166
speculative, speculation, 16, 43, 86, 106, 109, 111, 112, 120, *121–123*, 126, 148, 149–150, 152, 156, 158, 165, 187 n. 48, 199 n. 40, 204 n. 30, 208 n. 139
Speight, Allen, 16, 195–196 n. 14
Spinoza, Baruch, 86, 193 n. 55
Spivak, Gayatri Chakravorti, 107
Steinkraus, Warren E., 16
Stewart, Jon, 188 n. 11, 188 n. 13
Strauss, Richard, 41
Sussman, Henry, 201 n. 34

Taylor, Charles, 4, 41, 194 n. 61, 206 n. 92

Vater, Michael, 12

Vattimo, Gianni, 206 n. 92

Wagner, Richard, 143
Wicks, Robert, 16
Wood, Alan W., 209 n. 7
Wyss, Beat, 16

Yeats, William Butler, 164

Zeno (of Elea), 51
zero, 77, 78, 126
Žižek, Slavoj, 4, 18, 119, *136–145*, 146, 157, 159, 176, 209 n. 6, 210 n. 11

www.ingramcontent.com/pod-product-compliance
Lightning Source LLC
Chambersburg PA
CBHW051521230426
43668CB00012B/1690